FINANCIAL ASTROLOGY

Almanac 2022

Trading & Investing Using the Planets

M.G. Bucholtz, B.Sc, MBA, M.Sc.

A WOOD DRAGON BOOK

Financial Astrology Almanac 2022
Trading & Investing Using the Planets

Copyright © 2021 by M.G. Bucholtz

Published by :
Wood Dragon Books,
Box 429, Mossbank, Saskatchewan, Canada, S0H 3G0
http://www.wooddragonbooks.com

ISBN : 978-1-989078-82-2

To contact the author:
supercyclereport@gmail.com

DEDICATION

To the many traders and investors who at some visceral level suspect there is more to the financial market system than P/E ratios and analyst recommendations.

You are correct. There is more. Much more. Rooted in astronomical and astrological timing, the markets are a rich tapestry of interwoven cycles. This book will add a whole new dimension to your trading and investing activities.

DISCLAIMER

All material provided herein is based on material gleaned from mathematical and astrological publications researched by the author to supplement his own trading. This publication is written for those who actively trade and invest in the financial markets and who are looking to incorporate astrological phenomena and esoteric math into their market activity. While the material presented herein has proven reliable to the author in his personal trading and investing activity, there is no guarantee this material will continue to be reliable into the future.

TABLE OF CONTENTS

INTRODUCTION

Many traders and investors think company press releases, media news opinions, quarterly earnings reports, and analyst targets drive stock prices and major index movements.

I disagree. I believe the financial markets are defined by various cycles of planetary movement across time. Time, itself, can be expressed in fractal intervals over extremely long stretches of time related to Saros cycles and even to the Precession of the Equinox. Time can be expressed over shorter intervals from one New Moon to the next. Time can even be expressed by intervals between significant religious celebratory events. Overlap and interweave together these time intervals with cycles of planetary movement and the result will be the ups and downs of price that characterize a stock chart, a commodity price chart, or the chart of a major index. The average trader or investor who remains fixated on press releases and analyst opinions is unable to discern this rich tapestry of cyclical movement.

Cycles of planetary activity include the Jupiter/Saturn Gann Master cycle which unfolds over two decades of heliocentric planetary movement.

Interwoven with this long cycle is the 18.6-year movement of the North Node through the signs of the zodiac. This nodal cycle broadly defines overall economic activity. The repeated cyclical movement of planets above and below the ecliptic plane align to swing highs and swing lows on commodity futures. Irregular cycles of time between events of Mercury or Venus being retrograde often align to trend reversals on stocks and equity indices. Likewise for times between Mercury being at elongation extremes and Venus being at Superior and Inferior conjunctions. Cycles also arise from annual celebratory events delineated in the Hebrew Calendar and multi-year Shemitah events. Cycles arise from the Hebrew Alef Bet, and when expressed by heliocentric movement of Venus or Mars, can be seen aligning to market turning points. At the shorter end of the spectrum, the cycle from one New Moon to the next can also be seen to have a bearing on the market. Times of Moon being Void of Course also align to expressions of notable volatility on equity indices.

Even though the mainstream media refuses to embrace astrology or cycles of any sort as valid tools for timing the markets, there are powerful players in the major financial centres of the globe who *do* embrace astrology and its cycles. They use these cycles to their advantage to make money in an uptrending market. They also use these cycles to induce trend changes across markets. As the trend turns and markets start to fall, these players profit from their short positions while the average investor on the street experiences angst and sleepless nights knowing the markets are trending down and against them.

"How and why these various cycles have come to be?" is the burning question that remains unanswered. As I studied these cycles, I developed a new sense of awe for what I deem to be a higher power that guides the Universe. After reading this Almanac, you too may have reason to pause and ponder the power of the cosmos.

"Who are the powerful players that use astrology to move markets?" is another burning question. *Is it a select group at J.P. Morgan? Is it a group in a dark-panelled office in London?* I will likely never know. But what I do know is, that paying attention to planetary cycles is not a new concept. Ancient civilizations as far back as the Babylonians recognized cyclical activity, but in a more rudimentary form. Their high priests tracked and recorded changes in the emotions of the people. These diviners and seers also tracked events, both fortuitous and disastrous. Although they lacked the ability to fully comprehend the celestial mechanics of the planetary system, they were able to visually spot the planets Mercury, Venus, Mars, Jupiter, and Saturn in the heavens. They correlated changes in human emotion and changes in societal events to these planets. They assigned to these planets the names of the various deities revered by the people. They identified and named various star constellations in the heavens and divided the heavens into twelve signs. This was the birth of *astrology* as we know it today.

Stories of traders benefiting from astrology are also not new. In the early 1900s, esoteric thinkers such as the famous Wall Street trader W.D. Gann reportedly made massive gains when he realized that cycles of astrology bore a striking correlation to financial market action. Gann is most famous for identifying the Saturn/Jupiter cycle which he labelled the Gann Master Cycle. He followed cyclical activity of Jupiter and Neptune when trading wheat and corn futures. He also delved deep into esoteric math, notably square root math which led him to discover his Square of Nine. The concept of price squaring with time is also a Gann construct. Today many traders and investors attempt to emulate Gann but they do so in a linear fashion, looking for repetitive cycles on the calendar. What they are missing is the astrology component, which is anything but linear.

In the 1930s, Louise McWhirter contributed significantly to financial astrology. She identified an 18.6-year cyclical correlation between the general state of the American economy and the position of the North Node of the Moon in the zodiac. Her methodology extended to include the transiting Moon passing by key points of the 1792 natal birth horoscope of the New York Stock Exchange. She also identified a correlation between price movement of a stock and those times when transiting Sun, Mars, Jupiter and Saturn made hard aspects to the natal Sun position in the stock's natal birth (first trade) horoscope.[1]

The late 1940s saw further advancements in the field of financial astrology when astrologer Garth Allen (a.k.a. Donald Bradley) produced his *Siderograph Model* based on aspects between the various transiting planets. Each aspect as it occurs is given a sinusoidal weighting as the orb (separation) between the planets varies. Bradley's model is as powerful today as it was in the late 1940s. [2]

As the 1950s dawned, academics at institutions like Yale and Harvard came to dominate the discussion and the cycles of astrology were soon swept aside out of public view. Cyclical analysis was replaced by academic creations like *Modern Portfolio Theory* and the *Efficient Market Hypothesis*. These persisted for several decades until coming under severe scrutiny with the 2000 tech bubble meltdown and again with the 2009 sub-prime mortgage crisis which nearly derailed the global economy.

In the past decade, astrology has made a reappearance. The software designers at *Market Analyst/Optuma* now have an impressive financial astrology platform built into their charting program. More recently, author and trader Fabio Oreste published a book on Quantum Lines, a powerful tool based on the work of Einstein, Niels Bohr and Bernhard Reimann.

Think back to the dark days of late 2008 when there was genuine concern over the very survival of the financial market system. This timeframe was the end of an 18.6-year cycle of the North Node traveling around the zodiac. To those players at high levels in the financial system who understood astrology, this period was a prime opportunity to feast off the fear of the investing public and the anxiety of government officials who were standing at the ready with lucrative bailout packages. The market low in March 2009 came at a confluence of a Mars and a Neptune quantum point. Curiously enough, Mars and Neptune are deemed to be the planetary rulers of the New York Stock Exchange.

Think back to August 2015 and the market selloff that the financial media did not see coming. The reality is this selloff started at a confluence of two events: Venus retrograde event and the appearance of Venus as a morning star after having been only visible as an evening star for the previous 263 days.

Remember the early days of 2016 when Mercury was retrograde and the markets hit a rough patch? Remember the weakness of June 2016 when Venus emerged from conjunction to become visible as an Evening Star?

Do you recall the dire predictions for financial market calamity following the election of Donald Trump to the White House? When the markets instead powered higher, analysts were flummoxed. Venus was making its declination minima right at the time of the American election. Venus declination minima events bear a striking correlation to changes of trend on US equity markets.

What about the early days of 2018 when fear once again gripped the system? Venus was at its declination minimum. Markets reached a turning point in the first week of October 2018 when Venus was again at a declination low. Add the fact that Venus turned retrograde at the same time and the fear starts to make sense. Markets sold off sharply into mid-December before starting to recover. Sun was conjunct Saturn at this time which correlates strongly to trend changes on equity markets. The North Node had also just changed zodiac signs, an event which also aligns with trend changes.

Markets hit a sudden rough patch in early August 2019. Mercury had just finished a retrograde event and the Federal Reserve cut interest rates as Moon transited a key point on the NYSE 1792 natal horoscope.

US equity markets peaked in late February 2020 and went into total spasm in March 2020. Mercury was retrograde and Mars had just made its declination minimum. A powerful fractal cycle from a crisis event in January 1920 suggested another crisis event in March 2020.

Do you recall the confusion surrounding the 2020 US Presidential Election? The day of the election, Mercury finished a retrograde event and heliocentric Jupiter and Saturn were exactly at 0 degrees of separation. To have these two events occur right on the election date is a rarity. The events of this election will be hotly debated for years to come.

As I craft this manuscript, the markets have been rallying smartly for much of 2021 fueled by $120 billion in fiat liquidity being injected into the financial system each month. Any dips in the market were quickly erased as institutional money rushed in to buy, buy, buy. Inflection points aligning to astrology were largely smoothed over through much of 2021. I was pleased to see astrology re-assert itself and interrupt the uptrend in early September, 2021. This trend change, which delivered a 5% drawdown, aligned to a significant religious date in the Hebrew calendar.

And, so it goes. Cycles continue to unfold as time marches on. The Federal Reserve fiat liquidity that has cushioned the markets since March 2020 is going to be taken away in the first part of 2022. The markets will once again start to display cyclical volatility. Holding stocks for the long term while expecting the Federal Reserve to back-stop the markets will no longer be a valid strategy. People who view the markets through the lens of analyst opinions and media blather will be unable to see these cycles, which are hidden in plain view. They will ride an emotional roller coaster as their financial planners tell them that investing is for the long term and not to worry. On the other hand, investors who are able to identify these cycles will be able to take steps to protect themselves and profit accordingly.

I personally began to embrace financial astrology in 2012 which was a monumental shift given that my educational background comprises an Engineering degree, an MBA degree and a M.Sc. degree. My approach to astrology is thus heavily slanted towards mathematics and analytical science versus the classical approach.

This Almanac begins by offering the reader a look at the basic science of astrology. What then follows is an examination of the various cycles that I believe drive the performance of the financial markets. For each type of cycle, I delineate times in the calendar months of 2022 when investors ought to be alert to possible trend changes on the US equity markets. In this age of global connectedness, moves on the S&P 500 are often quickly reflected in other global indices. I provide a look at various commodity futures and the astro phenomena that influence their price action. I explain the concept of Price Square Time as well as the concept of Quantum Price Lines, a powerful esoteric technique that can be used when applying astrology to making trading and investing decisions.

When applying astrology to trading and investing, it is vital at all times to be aware of the price trend. The chart indicators developed by J. Welles Wilder are very effective at identifying trend changes. In particular, the Parabolic Stop & Reverse and the Volatility Stop are two indicators that should be taken seriously. As a trader and investor, look for a change of trend that aligns to an astrology event. When you see the trend change, you should take action. Whether that action is implementing a long position, a short position, an Options strategy, or just tightening up on a stop loss will depend on your personal appetite for risk and on your investment and trading objectives. Using astrology for financial investing is not about taking action at each and every astro event that comes along because not all astro events are powerful enough to induce a change of trend.

Author Note:

This Almanac, which is my eighth such annual publication, is designed to be a resource to help you stay abreast of the various astro events that 2022 holds in store. I am also the author of several other astrology books. In addition, I publish a bi-weekly subscription-based newsletter called The Astrology Letter. Through all of my written efforts, I hope to encourage people to embrace financial astrology as a valuable tool to aid in trading and investing decision making.

CHAPTER ONE

Fundamentals

The Sun is at the center of our solar system. The Earth, Moon, planets and various other asteroid bodies complete our planetary system. In addition to the Sun and Moon, there are eight celestial bodies important to the application of astrology to the financial markets. These planets are Mercury, Venus, Mars, Jupiter, Saturn, Uranus, Neptune, and Pluto. Figure 1 illustrates these various bodies and the spatial relation to the Sun on the ecliptic plane. Mercury is the closest to the Sun while Pluto is the farthest away.

The Ecliptic and the Zodiac

The various planets and other asteroid bodies rotate 360 degrees around the Sun following a path called the *ecliptic plane*. As shown in Figure 2, Earth (and its Equator) is slightly tilted (approximately 23.45 degrees) relative to the ecliptic plane.

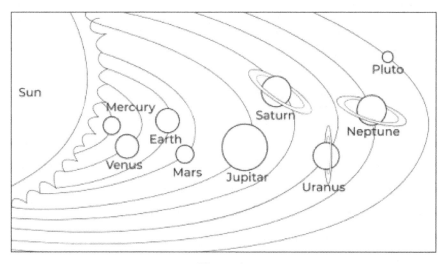

Figure 1
The Planets

Projecting the Earth's equator into space produces the *celestial equator plane*. There are two points of intersection between the ecliptic plane and celestial equator plane. Mathematically, this makes sense as two non-parallel planes must intersect at two points. These points are commonly called the *vernal equinox* (occurring at March 20th) and the *autumnal equinox* (occurring at September 20th). You will recognize these dates as the first day of Spring and the first day of Fall, respectively. Dividing the ecliptic plane into twelve equal sections of 30 degrees results in what astrologers call the *zodiac*. The twelve portions of the zodiac have names including Aries, Cancer, Leo and so on. Ancient civilizations looking skyward identified patterns of stars called constellations that align to these twelve zodiac divisions. If these names sound familiar, they should. You routinely see all twelve names in the daily horoscope section of your morning newspaper.

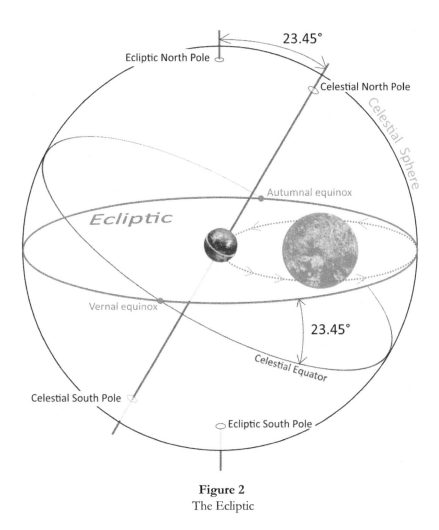

Figure 2
The Ecliptic

The Glyphs

Figure 3 illustrates the symbols that appear in the twelve segments of a zodiac wheel. The segments are more properly called signs. These symbols are called glyphs.

The starting point or zero degree point of the zodiac wheel is the sign Aries, located at the vernal equinox of each year. The vernal equinox is when, from our vantage point on Earth, the Sun appears at zero

11

degrees Aries. The autumnal equinox is when, from our vantage point on Earth, the Sun appears at 180 degrees from zero Aries (0 degrees of Libra).

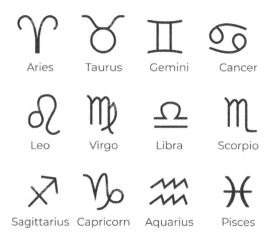

Figure 3
The Zodiac Wheel

The various planets are also denoted by glyphs, as shown in Figure 4.

Sun	☉	Saturn	♄
Moon	☽	Neptune	♅
Mercury	☿	Uranus	♆
Venus	♀	Pluto	♇
Mars	♂	Chiron	⚷
Jupitar	♃	Nodes	☊☋

Figure 4
The Glyphs

Declination

As the various celestial bodies make their respective journeys around the Sun, they can be seen to move above and below the celestial equator plane. This movement is termed declination. Celestial bodies experience declinations of up to about 25 degrees above and below the celestial equator plane.

Mercury, Venus, and Mars endure frequent changes in declination due to the gravitational force of the Sun. Planets like Jupiter, Saturn, Neptune, Uranus and Pluto also experience declination changes but these changes are slower to evolve. As this Almanac will illustrate, changes in the declination of a celestial body (most notably Mars and Venus) can affect the financial markets.

Parallel and Contra-Parallel

Declination can be viewed one planet at a time or by pairs of planets. Let's suppose that at a particular time Mars can be seen as being 10 degrees of declination above the celestial equator and at that same time Venus is at 9 degrees of declination. Let's further suppose that we allow for up to 1.5 degrees tolerance in our measurement of declinations. We would say these two planets were at parallel declination. Let's take another example and suppose that at a given time Jupiter was at 5 degrees of declination above the celestial equator and at that same time period Pluto was at 6 degrees declination below the celestial equator. Again, let's allow for up to 1.5 degrees of tolerance. We would say that Jupiter and Pluto were at contra-parallel declination. As this Almanac will show, parallel and contra-parallel events can influence the financial markets.

The Moon

Just as the planets orbit 360 degrees around the Sun, the Moon orbits 360 degrees around the Earth. The Moon orbits the Earth in a plane of motion called the lunar orbit plane. This plane is inclined at about 5 degrees to the ecliptic plane as Figure 5 shows. The Moon orbits Earth with a slightly elliptical pattern in approximately 27.3 days, relative to an observer located on a fixed frame of reference such as the Sun. This time period is known as a sidereal month. However, during one sidereal month, an observer located on Earth (a moving frame of reference) will revolve part way around the Sun. Because of this added movement, the Earth-bound observer will see a complete orbit of the Moon around the Earth in approximately 29.5 days. This 29.5-day period of time is known as a synodic month or more commonly a lunar month. The lunar month plays a key role in applying astrology to the financial markets.

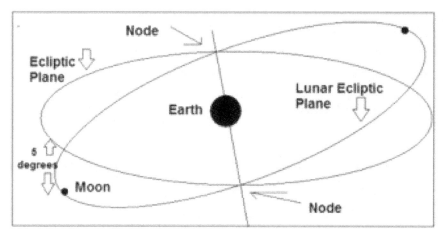

Figure 5
Lunar Orbit

The Nodes

A mathematical construct related to the Moon and central to financial astrology is the Nodes. The Nodes are the points of intersection between the Earth's ecliptic plane and the Moon's ecliptic plane. Figure 5 also illustrates the Nodes. In astrology, typically only the North Node is referred to. The North Node forms the basis for the McWhirter Method which will be discussed in Chapter 3.

Ascendant, Descendant, MC and IC

As the Earth rotates on its axis once in every 24 hours, an observer situated on Earth will detect an apparent motion of the constellation stars that define the zodiac. To better define this motion, astrologers apply four cardinal points to the zodiac, almost like the north, south, east and west points on a compass. These cardinal points divide the zodiac into four quadrants. The east point is termed the Ascendant and is often abbreviated Asc. The west point is termed the Descendant and is often abbreviated Dsc. The south point is termed the Mid-Heaven (Latin for Medium Coeli) and is often abbreviated MC or MH. The north point is termed the Imum Coeli (Latin for bottom of the sky) and is abbreviated IC.

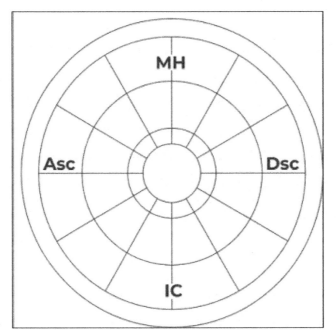

Figure 6
Cardinal Points

Figure 6 illustrates the placement of these cardinal points on a typical zodiac wheel. The importance of the Ascendant and Mid-Heaven will be emphasized in more detail when the McWhirter Method is discussed.

Geocentric and Heliocentric Astrology

Astrology comes in two distinct varieties – geocentric and heliocentric.

In geocentric astrology, the Earth is the vantage point for observing the planets as they pass through the signs of the zodiac. Owing to the different times for the planets to each orbit the Sun, an astrologer situated on Earth would see the planets making distinct angles (called aspects) with one another and also with the Sun. The aspects that are commonly used in astrology are 0, 30, 45, 60, 90, 120, 150 and 180 degrees. In financial astrology, it is common to refer to only the 0, 90, 120 and 180 degree aspects.

In heliocentric astrology, the Sun is the vantage point for observing the planets as they pass through the signs of the zodiac. An observer positioned on the Sun would also see the orbiting planets making aspects with one another.

To identify these aspects, astrologers use Ephemeris tables. For geocentric astrology, the *New American Ephemeris for the 21ˢᵗ Century* is commonly used. For heliocentric astrology, the *American Heliocentric Ephemeris* is a good resource.

For faster aspect determination, two excellent software programs available are *Millenium Trax* produced by AIR Software and *Solar Fire Gold* produced by software company Astrolabe. My preference is the Solar Fire Gold product. I also use a market platform called *Optuma/Market Analyst*. This brilliant piece of software, originally developed in Australia, allows the user to generate end of day price charts for equities and commodities from a multitude of exchanges and then overlay various astrological aspects and occurrences onto the chart. As your journey into astrology deepens, you might be tempted to spend the money to acquire a software program.

Synodic and Sidereal

The vantage point of either Earth or Sun leads to two more concepts, synodic and sidereal. These descriptors were discussed earlier in the context of the Moon. To an earth-bound observer, a synodic time period is the time between two successive planetary occurrences. That is, how many days does it take for Sun passing Pluto on the zodiac wheel to again pass Pluto? To a Sun-bound observer (a fixed frame of reference), a sidereal time period is the number of days (or years) it takes for a planet to orbit the Sun. The sidereal times form the basis for heliocentric astrology. The data in Figure 7 presents synodic and sidereal data.

PLANET	SYNODIC PERIOD	SIDEREAL PERIOD
Mercury	116 days	88 days
Venus	584 days	225 days
Mars	780 days	1.9 years
Jupiter	399 days	11.9 years
Saturn	378 days	29.5 years
Uranus	370 days	84 years
Neptune	368 days	164.8 years
Pluto	367 days	248.5 years

Figure 7
Synodic and Sidereal Data

Retrograde

Think of the planets orbiting the Sun as a group of cars travelling around a racetrack. Consider what happens as a fast moving car approaches a slower moving car from behind. At first, all appears normal. An observer in the fast moving car sees the slower moving car heading in the same direction. Gradually, the observer in the fast car sees that he will soon overtake the slow car. For a brief moment in time as the fast car overtakes the slower car the observer in the fast car notices that the slower car appears to stand still and even move backwards. Of course the slow car is not really standing still. This is simply an optical illusion.

These brief illusory periods are what astrologers call retrograde events. To ancient societies, retrograde events were of great significance as human emotion was often seen to be changeable at these events. Is it possible that our DNA is hard-wired such that we feel uncomfortable at retrograde events?

From the vantage point of an observer on Earth, there will be three or four times during a year when Earth and Mercury pass by each other on this celestial racetrack. There will be one or perhaps two times per year when Earth and Venus pass each other. There will be one time every two years when Earth and Mars pass each other. Retrograde events all too often will see a short term trend change develop.

Elongation and Conjunction

From an observer's vantage point on Earth, there will also be times when planets are seen to be at maximum angles of separation from the Sun. These events are what astronomers refer to as maximum easterly and maximum westerly elongations. These events definitely have a correlation trend changes on the markets.

Mercury and Venus are closer to the Sun than is the Earth. From our vantage point on Earth, there will be times when Mercury and Venus are between the Earth and the Sun. Likewise, there will be times when the Sun is between the Earth and Mercury or Venus. On the zodiac wheel, the times when Mercury or Venus are in the same zodiac sign and degree as the Earth are what astronomers call conjunctions.

An Inferior Conjunction occurs when Mercury or Venus is between Earth and the Sun.

A Superior Conjunction occurs when the Sun is between Earth and Mercury or Venus. Figure 8 illustrates the concept of elongation and conjunction.

Conjunction events occur on either side of retrograde events. For example, in 2018 Venus was retrograde from October 5 to November 15. Its actual Inferior Conjunction was recorded on October 26.

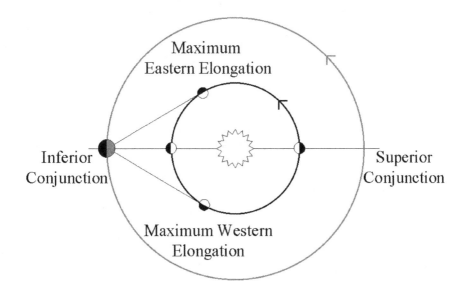

Figure 8
Superior and Inferior Conjunction

After Venus has been at Inferior Conjunction, it will be visible as a Morning Star.

After it has been at Superior Conjunction, it will be visible as an Evening Star.

Venus was at Superior Conjunction on March 28, 2013 (8 Aries), October 25, 2014 (1 Scorpio), June 6, 2016 (16 Gemini), and January 8, 2018 (18 Capricorn). Venus was at Inferior Conjunction on June 6, 2012 (15 Gemini), January 11, 2014 (21 Capricorn), August 15, 2015 (22 Leo), March 25, 2017 (4 Aries), and October 26, 2018 (3 Scorpio). If you plot these groups of Superior Conjunction events on a zodiac wheel, you will note they can be joined to form a 5-pointed star called a pentagram. Likewise, the Inferior Conjunction events can be plotted and joined to form a pentagram. Such is the elegance and mystique of the cosmos.

CHAPTER TWO

The Gann Master Cycle

W.D. Gann closely followed the cycles of Jupiter and Saturn. To an observer situated on the fixed vantage point of the Sun, Jupiter would be seen orbiting the Sun in about 12 years and Saturn in just over 29 years. Gann interpreted these orbital cycles one step further and noted that every 20 or so years heliocentric Jupiter and Saturn were at conjunction, in a particular sign of the zodiac, separated by zero degrees. This is what he called the *Master Cycle*.

Gann further noted that the financial markets were affected by this Master Cycle. The market crash of 1901 aligned to a conjunction of these two outer planets. In 1920, the U.S. economy encountered a recession, and in 1921 the financial markets reached a low point again at the conjunction of these two heavy-weight planets. In 1941, the markets recorded a low point a couple months after another Jupiter-Saturn conjunction. The early-1960s should have seen another major turning point at the end of a Master Cycle. There was a drawdown in the U.S. equity markets that historians now call the Kennedy Slide, but it did not align perfectly to the Master Cycle. As it turns out, this one exception to the Master Cycle was affected by the far more powerful *Cardinal Cross.*

Picture a rectangle with its parallel sides and parallel ends. Now, place this rectangle inside of the zodiac. The corner points of the rectangle are pairs of planets. This 1965 Cardinal Cross involved eight celestial bodies (Neptune was the odd planet out). With the Dow Jones Average at near 8000 points, the Cardinal Cross marked a significant turning point for the US equity market. It would not be until 1995 when the US equity market would again test the 8000 level.

In the Spring of 1981, Jupiter and Saturn recorded a Master Cycle conjunct event. A handful of months later, the U.S. equity markets recorded a very important low that marked the onset of a massive bull market that ran until the next conjunction event in mid-2000. If you were active in the markets in 2000, you may have painful memories of the tech bubble and of money lost in your brokerage account. Figure 9 illustrates how the S&P 500 was making its peak as Jupiter and Saturn were making their conjunction.

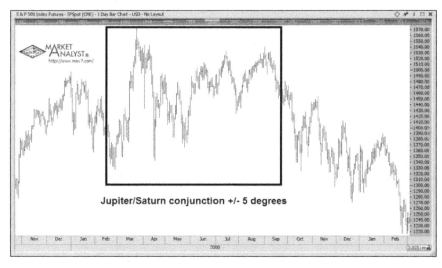

Figure 9
Jupiter / Saturn Conjunction in 2000

What happened after this conjunction is even more interesting. The market ground painfully lower until late 2002. Rallies along the way were systematically crushed.

In November 2020, heliocentric Saturn and Jupiter again made their zero-degree conjunction right at the time of the US Presidential election. As Figure 10 illustrates, the markets appear to be behaving differently than at the start of the last Master Cycle in 2000. So far, this Master Cycle has had a positive tone to it. But the Federal Reserve has been injecting $120 billion per month of liquidity into the financial system in response to the Covid crisis. As the Fed tapers back this liquidity starting in early 2022 the true strength of the market will be revealed. Factors at play include the global supply chain being upended; the Chinese Communist Party tightening its grip on society and on listed companies; interest rates are creeping higher and inflation is looming. All of these factors could have less than desirable effects on earnings heading into 2022.

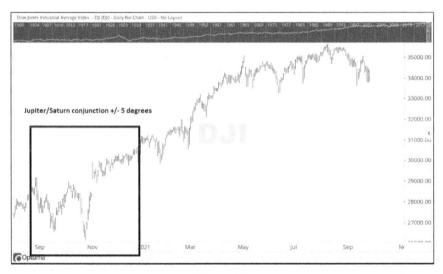

Figure 10
Jupiter / Saturn Conjunction in 2020

Moreover, a Shemitah year has now commenced as of September 2021 and will remain in effect until September 2022. Shemitah years are notorious for unexpected market moves. I am curious to see what market sectors become favorable as this Master Cycle unfolds. Will Gold finally gallop higher as precious metals pundits have been calling for? Will crypto currencies become more mainstream? Will the US Dollar cease to be the global reserve currency? Will the conventional oil industry diminish further? What new energy technologies will emerge? Has the Covid pandemic permanently altered the notion of work? The coming year will shed light on these and many other questions.

CHAPTER THREE

The 18.6 Year Cycle

In addition to the Gann Master Cycle, there exists another longer cycle that more people ought to be aware of.

This cycle was first written about in the 1930s by a mysterious figure called Louise McWhirter.[1] I say *mysterious* because in all my research I have neither come across any other writings by her nor have I found reference to her in other manuscripts. I am almost of the opinion that the name was a pseudonym for someone seeking to disseminate astrological ideas while remaining anonymous.

The Moon orbits the Earth in a plane of motion called the lunar ecliptic. Two planes that are not parallel will always intersect at two points. The two points where the lunar ecliptic intersects that plane of motion of planet Earth are termed the North Node and South Node. McWhirter recognized that the transit of the North Node of the Moon around the zodiac wheel takes 18.6 years and that the Node progresses in a backwards motion through the zodiac signs.

Through examination of copious amounts of economic data provided by Leonard P. Ayers of the Cleveland Trust Company, McWhirter was able to conclude that when the North Node moves through certain zodiac signs, the economic business cycle reaches a low point and when the Node is in certain other signs, the business cycle is at its strongest.

This line of thinking is still with us today. A notable authority embracing this cycle is Australian economist Fred Harrison. In his published works, he discusses this long economic cycle going back to the Industrial Revolution. But, to maintain respect in academia, he stops just shy of stating a connection to astrology.

McWhirter was able to discern the following from the Cleveland Trust data:

- ✪ As the Node enters Aquarius, the low point of economic activity is reached.

- ✪ As the Node leaves Aquarius and begins to transit through Capricorn and Sagittarius, the economy starts to return to normal.

- ✪ As the Node passes through Scorpio and Libra, the economy is functioning above normal.

- ✪ As the Node transits through Leo, the high point in economic activity is reached.

- ✪ As the Node transits through Cancer and Gemini, the economy is easing back towards normal.

- ✪ As the Node enters the sign of Taurus, the economy begins to slow.

- ✪ As the Node enters Aquarius, the low point of economic activity is reached and a full 18.6 year cycle is completed.

McWhirter further observed some secondary factors that could influence the tenor of economic activity in a good way, no matter which sign the Node was in at the time:

- ✿ Jupiter being 0 degrees conjunct to the Node
- ✿ Jupiter being in Gemini or Cancer
- ✿ Pluto being at a favorable aspect to the Node

McWhirter also observed some secondary factors that can influence the tenor of economic activity in a bad way, no matter which sign the Node was in at the time:

- ✿ Saturn being 0, 90 or 180 degrees to the Node
- ✿ Saturn in Gemini or Cancer
- ✿ Uranus in Gemini
- ✿ Uranus being 0, 90 or 180 degrees to the Node
- ✿ Pluto being at an unfavorable aspect to the Node.

In the Summer of 2019, there were media rumblings about an imminent recession. But, several large investment banking firms in New York weighed in with a collective dissenting opinion that pushed talk of recession off into the future. I perceive this to be evidence that the large money firms in New York are paying close attention to the 18.6-year cycle. Soon enough, the obedient media talking heads were echoing this non-recessionary sentiment.

In early 2020, North Node was in the sign of Cancer. The economy was gently easing, in alignment with McWhirter's work. None of the above mentioned secondary bad factors were in play. Then suddenly global financial markets were turned upside down by a virus emanating from Wuhan, China. The fact that this economic wallop was not forecast by McWhirter astrology opens the door to all sorts of conspiratorial speculation that is well beyond the tenor of this book.

As I pen this manuscript, the Node is working its way through Gemini and will enter Taurus in early 2022. As the Node moves through Taurus in 2022, the prospect of recession and slowdown will loom ever larger.

Recession will manifest in a big way starting in late 2024. By 2026, the Node will be in Aquarius to mark the end of the 18.6-year cycle and very likely the depths of another financial crisis.

In 2026, Uranus will be in Gemini and also at a hard 90 degrees to the Node. This astro positioning warns of a very negative time. In fact, it warns of possible war. The 1776 natal horoscope for the USA has Uranus in the sign of Gemini. By mid-2027 Uranus will be exactly conjunct to the 1776 Uranus natal position at 8 of Gemini. Uranus takes 84 years to travel one time through the zodiac. Subtract Uranus cycles from the year 2027 and past dates aligning to World War II, the US Civil War, and the War of Independence all come into focus.

But these events are still a good four years distant. Be aware of what is ahead and plan accordingly.

CHAPTER FOUR

Fractal Cycles

It is important for traders and investors to understand that the study of fractal math centers around iterations of a basic shape. Consider the basic shape shown in Figure 11. In Figure 12, this basic pattern (the seed pattern) has been continued through many iterations. It soon becomes apparent that the iterated pattern takes on the appearance of a branched tree.

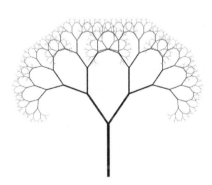

Figure 12

Figure 11

In the financial markets the starting point for a repetitive fractal can be taken from a wide variety of seed events. Late in 2021, I happened upon a book by author Gregg Braden (*Fractal Time*) at a used bookstore.[1]

Braden's starting point for fractal analysis is the Precession of the Equinox. Each year in the 3rd week of March, we celebrate the Spring Equinox. The precise time of the Equinox coincides with the declination of the Sun being at 0 degrees above the ecliptic. Currently, the Spring Equinox has the Sun in the sign of Aries. Every 71 years, the Equinox will occur 1 degree earlier. After 25,878 years the Equinox will again be back at its start point. This backwards movement is the Precession of the Equinox. Braden divides the 25,878-year period into 5 intervals of 5175 years each. Braden argues that we are currently in the midst of one of these 5175-year periods which began in 3114 BC. The period will end in 2061.

Using Braden's method, one would take a year that is of curious interest (the seed event) and add to it the value of 3114. One would then divide this sum by 5175 and multiply the result by phi (0.618). The result of this multiplication will next be multiplied by the remaining years in the cycle.

For example, in January 1819, there was a financial panic in the US. Author Clyde Haulman[2] writing in the journal Financial History describes that in the early 1800s many individual states in the western part of America had issued their own bank notes. The eastern part of the US was using its own bank notes. The notes from the western states started depreciating in value around 1816. Holders of these western notes attempting to redeem them for cash at eastern banks were soon stopped by the US Treasury Department. This monetary contraction set the stage for an eventual panic. Haulman writes:

With a monetary contraction underway, the continued retirement of federal debt, much of it to foreigners, and declines in the overseas markets for American staples, the United States economy was headed for disaster. The most dramatic aspect of the disaster was a rapid deflation as prices fell 30.6 percent between 1818 and 1821. Stagnation of real output that for some parts of the country lasted well into the 1820s. Real GNP fell in 1819 and was flat over the period 1818–1821. The young republic's rude introduction to boom-and-bust capitalism reported by these sources was a complex combination of financial market volatility, swings in international market demand, and federal government financial activity.

January 1819 seems to have been the worst of the crisis as bank credit plunged.

Applying Braden's method, January is $1/12^{th}$ of the calendar year. In decimal form, $1/12^{th}$ is 0.083.

1819.083 + 3114 = 4933.083
(4933.083/5175) * phi = 0.58911
5175 – 4933.083 = 241.917
241.917 * 0.58911 = 142.515
4933.083 + 142.515 = 5075.59
5075-3114 = 1961.598 = July 1961.

These calculations suggest that the next fractal event stemming from the 1819 event would occur in mid-1961.

A look back at history shows that at this time in 1961 the market was reaching a peak. A handful of months later the market starting selling off. The damage ended with a drawdown of 22%. This event has been dubbed the Kennedy Slide. Stocks has become too expensive and a return to more sensible levels was overdue.

Taking the 1961.598 figure and running it through the above process yields a value of 2021.848. In round figures this takes us to October 2021. Fractal math is thus suggesting market weakness will start in October 2021. As I finish this manuscript, the equity market is displaying some weak behavior in accordance with what the fractal calculations are suggesting. The conditions are no doubt ripe for a serious drawdown in the financial system. But the Federal Reserve is acting as a backstop. As the S&P 500 approached a 5% drawdown, fiat money liquidity entered the system and the S&P 500 reversed its losses. By not allowing the market to fully react in accordance with fractal time, the crisis event is merely being pushed forward in time. The adage "forewarned is forearmed" certainly applies to fractal cycles. By being aware that a fractal cycle event is pending, traders and investors can at least move to protect themselves should the crisis event unfold.

As another example, there was a financial panic event in 1893. Richardson and Sablik[3] of the Federal Reserve Bank of Richmond wrote:

The Panic of 1893 was one of the most severe financial crises in the history of the United States. The crisis started with banks in the interior of the country. Instability arose for two key reasons. First, gold reserves maintained by the U.S. Treasury fell to about $100 million from $190 million in 1890. At the time, the United States was on the gold standard, which meant that notes issued by the Treasury could be redeemed for a fixed amount of gold. The falling gold reserves raised concerns at home and abroad that the United States might be forced to suspend the convertibility of notes, which may have prompted depositors to withdraw bank notes and convert their wealth into gold. The second source of this instability was that economic activity slowed prior to the panic. The recession raised rates of defaults on loans, which weakened banks' balance sheets. Fearing for the safety of their deposits, men and women began to withdraw funds from banks. Fear spread and withdrawals accelerated, leading to widespread runs on banks.

In June, bank runs swept through midwestern and western cities such as Chicago and Los Angeles. More than one-hundred banks suspended operations. From mid-July to mid-August, the panic intensified, with 340 banks suspending operations. As these banks came under pressure, they withdrew funds that they kept on deposit in banks in New York City. Those banks soon felt strained. To satisfy withdrawal requests, money center banks began selling assets. During the fire sale, asset prices plummeted, which threatened the solvency of the entire banking system. In early August, New York banks sought to save themselves by slowing the outflow of currency to the rest of the country. The result was that in the interior local banks were unable to meet currency demand, and many failed. Commerce and industry contracted. In many places, individuals, firms, and financial institutions began to use temporary expediencies, such as scrip or clearing-house certificates, to make payments when the banking system failed to function effectively.

From this description of events, it appears that September was perhaps the bottom of the crisis. September 1893 equates to 1893.75

1893.75 + 3114 = 5007.75
(5007.75/5175) * phi = 0.5980
5175 – 5007.75 = 167.25
167.25 * 0.5980 = 100.01
5007.75 + 100.01 = 5107.76
5107.76-3114 = 1993.76 = late September 1993.

Late 1993 witnessed the signing of NAFTA, the free trade agreement between Canada, Mexico, and the US. This agreement set in motion the events for what would turn into a crisis in 1994. Foreign investors soon expressed a willingness to lend money to Mexico in exchange for US-dollar denominated bonds. A series of destabilizing political events in 1994 caused investors to demand a greater return on bond purchases. The Mexican government acceded to these requests which caused an appreciation in the value of the peso. The stronger currency quickly led to a serious trade deficit and a flight of capital out of the country once investors realized the peso was overvalued. In late 1994, the government abruptly devalued the peso which in turn posed risks to payments on US-dollar denominated debt. The US government ultimately ended up providing Mexico with a $50 billion rescue package. The S&P 500 fell 1.5%, the first time since 1990 that it had fallen.[4]

Taking the 1993.76 figure and passing it through the algorithm, yields 2034.77 or September 2034 for the seed of another crisis event.

1907 saw another financial panic. As Moen and Tallman[5] of the Federal Reserve Bank of Atlanta describe:

The Panic of 1907 was the first worldwide financial crisis of the twentieth century. It transformed a recession into a contraction surpassed in severity only by the Great Depression.

In reading their synopsis of the crisis, it appears that November 1907 was the worst of the crisis. November 1907 equates to 1907.91.

1907.91 + 3114 = 5021.91
(5021.91/5175) * phi = 0.5997
5175 – 5021.91 = 153.09
153.09 * 0.5997 = 91.8
5021.91 + 91.8 = 5113.71
5113.71-3114 = 1999.71 = late August 1999.

At this 1907 timeframe, the seeds were sown for what would be an event in mid-1999. By mid-1999, the tech bubble was inflating in a dangerous fashion. It was just going to be a matter of time until it all fell apart.

In late December 1924, the Dow Jones Average ramped higher surpassed its previous high made in late 1916. Using Braden's math, this equates to an event in 2007. Late 2007 marked a peak in the equity markets and the start of what would be the sub-prime mortgage crisis. Just as 1924 signaled the start of a wild excursion higher on the market, 2007 signaled the start of a wild excursion lower on the market.

And, so it goes. Find a crisis event, use Braden's fractal math and one can arrive at the timing of a future crisis. In other words, just as the tree-like image in Figure 12 started from a basic shape, financial crises across time have all started from a seed event in the annals of time.

I will leave you with one final example which deviates from Braden's technique of using a portion of the Equinox Precession as the starting pattern. Consider instead a Saros series of eclipses. As the NASA Eclipse website notes:

Any two eclipses separated by one Saros cycle share very similar geometries. They occur at the same node with the Moon at nearly the same distance from Earth and at the same time of year.

A Saros series comprises a number of Saros cycles. As NASA describes:

A typical Saros series for a solar eclipse begins when new Moon occurs about 18° east of either the North or South Node. After ten or eleven Saros cycles (about 200 years), the first eclipse will occur near the south pole of Earth. Over the course of the next 950 years, each will be displaced northward by an average of about 300 km. The last eclipse of the series occurs near the north pole of Earth. The next approximately ten eclipses will be partial events with successively smaller magnitudes. Finally, the Saros series will end a dozen or more centuries after it began at the opposite pole. Due to the ellipticity of the orbits of Earth and the Moon, the exact duration and number of eclipses in a complete Saros is not constant. A series may last 1226 to 1550 years and is comprised of 69 to 87 eclipses.[6]

For example, Saros series 150 started on August 24, 1729 and will conclude at September 29, 2991 (1262 years). If one takes the fractal seed event as being the start of the recession in January 1920 [7] [8] and applies Braden's math, the calculation says a future crisis event will occur in 100.17 years from January 1920. That equates to March 2020. I think we all know the significance of that Covid-related date.

CHAPTER FIVE

Venus Cycles

Cycles of Venus play a key role in financial astrology.

Venus orbits the Sun in 225 days relative to an observer standing at a fixed venue like the Sun.

To an observer situated on Earth (a moving frame of reference), Venus appears to take 584 days to orbit the Sun.

During this 584-day period, the Earth-bound observer will note periods of time when Venus is not visible in the early morning nor in the evening sky. This is because the planet is between Earth and Sun in its orbital journey. This is the Inferior Conjunction. As Venus slowly moves out of this conjunction, it will become visible as the Morning Star. During that part of its journey when Venus is 180 degrees opposite Earth, it is said to be at Superior Conjunction. As it moves out of this conjunction, it becomes visible as the Evening Star.

Venus was at Superior Conjunction on March 28, 2013 (8 Aries),

October 25, 2014 (1 Scorpio), June 6, 2016 (16 Gemini) and January 8, 2018 (18 Capricorn). Venus was at Inferior Conjunction on June 6, 2012 (15 Gemini), January 11, 2014 (21 Capricorn), August 15, 2015 (22 Leo), March 25, 2017 (4 Aries), and October 26, 2018 (3 Scorpio). If you plot these groups of Superior Conjunction events on a zodiac wheel you will note how they can be joined to form a 5-pointed star called a pentagram. Likewise, the Inferior Conjunction events can be plotted and joined to form a pentagram (5-pointed star). Such are the mysteries of our cosmos.

As Venus orbits around the Sun following the ecliptic plane, it moves above and below the plane. The high points and low points made during this travel are termed declination maxima and minima.

In early 2018, a Venus Superior Conjunction and declination minimum was followed closely by a steep 300 point sell-off on the S&P 500. A significant trend change immediately preceded an Inferior Conjunction and declination minimum in October 2018. This was followed by an acceleration of trend to the downside.

In July 2019, talk of a recession and an impasse with China over trade disputes caused a sharp sell-off on the S&P 500. But this sell-off event was nothing more than a Venus Superior Conjunction combined with a Venus declination maximum.

A Venus Superior Conjunction in early December 2019 failed to deliver a meaningful reaction. The Federal Reserve was busy ramping up liquidity to the banking system which had the effect of dampening the astrology cycles.

In early May 2020, Venus made a declination maximum and the S&P 500 exhibited a short, sharp 200-point drop. The damage could have been worse in the absence of the Federal Reserve's liquidity injection.

The Inferior Conjunction event on June 3, 2020 delivered a 200 point drop to the S&P 500. But again, the Federal Reserve came to the rescue with more fiat liquidity.

The days immediately following the January 14, 2021 Venus declination low were met with the S&P 500 dropping just 200 points (high to low).

The Superior Conjunction event of March 26, 2021 was immediately preceded by the Dow Jones dropping 1100 points (high to low) over six trading sessions.

The Venus declination maximum on June 4, 2021 was immediately followed by a 1500-point (high to low) drawdown on the Dow Jones Average.

For 2022,

☻ Venus will be at declination maximum on July 23

☻ Venus will be at minimum declination on December 13

☻ Inferior Conjunction will occur January 8, 2022

☻ Superior Conjunction will occur October 22, 2022.

As each of these Venus events approach, it is highly advisable to be alert for sudden market moves, higher or lower, which could impact your investment portfolio. As we enter 2022, the Federal Reserve will be tapering its stimulus. The buy-the-dip mentality might be severely challenged by events such as Venus conjunction or declination. To discern the effects of Venus on the markets watch the S&P 500 relative to its 21 day or 34 day moving average for evidence of a trend change. Also, watch a trend indicator (such as the Wilder Volatility Stop) giving evidence of a trend change.

Another cyclical event pertaining to Venus is its retrograde events. When discussing the basics of astrology in a previous chapter, I used the analogy of cars on a racetrack passing each other to explain retrograde. To further help understand the science of Venus retrograde, consider the diagram in Figure 13.

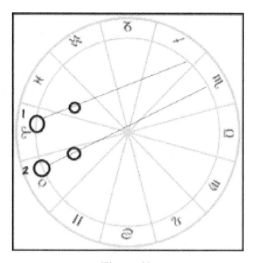

Figure 13
The concept of Venus retrograde

In 30 days of time, planet Earth (shown as the larger circles in the diagram) will travel 30 degrees of the zodiac (from point 1 to point 2).

But Venus is a faster mover. In the same 30 days of time, Venus (shown as the smaller circles) will travel through about 42 degrees of the zodiac, passing by Earth in the process. From our vantage point here on Earth, as Venus is setting up to pass Earth, we see Venus in the sign of Sagittarius. As Venus completes its trip past Earth, we see it in the sign of Scorpio.

It appears to have moved backwards in zodiac sign as it has passed by Earth. This is the concept of retrograde. To the ancients who did not fully understand how the cosmos worked, it must have been awe-inspiring to see a planet move backwards in the heavens relative to their constellation stars.

There is a curiously strong correlation between equity markets and Venus retrograde. Sometimes Venus retrograde events encompass a sharp market inflection point. Sometimes a market peak or bottom will follow closely behind a retrograde event. Sometimes a peak or bottom will immediately precede a retrograde event. When you know a Venus retrograde event is approaching, use a suitable chart technical indicator such as DMI or Wilder Volatility Stop to determine if the price trend is changing.

Figure 14 illustrates price behavior of the S&P 500 Index during May 2020 as Venus was retrograde.

Figure 14
S&P 500 Index and Venus retrograde

Knowing that the potential exists for sizeable moves, aggressive traders can avail themselves of these retrograde correlations. Less aggressive investors may simply wish to place a stop loss order under their positions to guard against sharp price pullbacks.

As a further example, Figure 15 illustrates price performance of Gold futures. Gold has a peculiar tendency to commence notable price moves during Venus retrograde events. I have examined this tendency going back to 1980 and have noted an 80% correlation with varying degrees of price gain. The May 2020 Venus retrograde event delivered a move higher by nearly $400 per ounce.

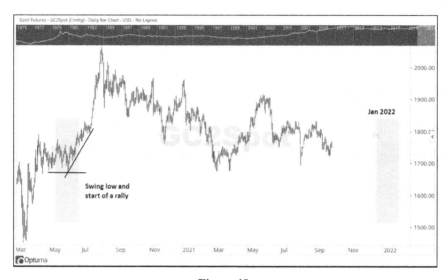

Figure 15
Venus retrograde and Gold

CHAPTER SIX
Mercury Cycles

Mercury is the smallest planet in our solar system. It is also the closest planet to the Sun. As a result of its proximity to the powerful gravitational pull of the Sun, Mercury moves very quickly, completing one heliocentric cycle of the Sun in 88 days.

Scientists at NASA have concluded that Mercury has a di-polar magnetic field. This field is thought to be about 1% as strong as Earth's magnetic field. Solar wind particles that have emanated from Sun coronal mass ejections are distorted and deflected by Mercury's magnetic field. In so doing, Mercury takes some of the burden off of Earth's magnetic field.

Scientists have also determined that Mercury has an eccentric orbit in which its distance from the Sun will range from 46 million kms to 70 million kms. When Mercury is nearer to the Sun (46 million kms away), it is moving at its fastest (56.6 kms per second). When Mercury is farther from the Sun (70 million kms away), it is moving slower (38.7 kms per second). [1]

Mercury at its closest orbital point to the Sun is called Mercury Perihelion. Mercury at its farthest orbital point to the Sun is called Mercury Aphelion.

In 2020, Mercury was at Perihelion just as the equity markets were reaching a peak ahead of the Covid panic selloff. The selloff lows came just as Mercury was nearing Aphelion. Mercury Perihelion and Aphelion events do not seamlessly correlate to all market inflection points. However, these dates should be anticipated by traders and investors nonetheless.

For 2022, Mercury will be at Perihelion at: January 16, April 14, July 11, and October 7. Aphelion dates will be: March 1, May 28, August 24, and November 20.

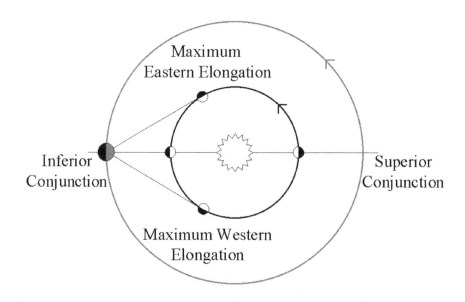

Figure 16
Elongation of Mercury

Related to Mercury's orbit is its elongation. As discussed earlier, elongation refers to the angle between a planet and the Sun, using Earth as a reference point. Figure 16 illustrates the notion of elongation.

In 2021,

☿ Mercury was at its greatest easterly elongation January 23, May 17, and September 13

☿ Greatest westerly elongation events occurred March 6, July 4, and October 25.

Figure 17 illustrates some of these 2021 events overlaid on a chart of the E-mini S&P 500. If one studies this chart, the argument can be made that these elongation events have a correlation to swing highs and swing lows. I first noticed this correlation in 2012 when I started studying the markets in the context of astrological cycles. This correlation extends well beyond just US equity futures. Wherever you live and whatever stocks, commodities or indices you follow, be alert to Mercury and its elongation events. I fervently believe these elongation events are a brilliant example of powerful players using astrology to move the markets when they desire.

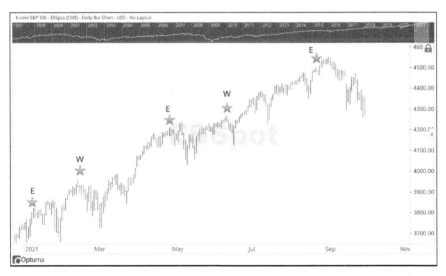

Figure 17
2021 Mercury Elongation events

For 2022,

- ☿ Mercury will be at its greatest easterly elongation January 7, April 29, August 27, and December 21

- ☿ Mercury will be at its greatest westerly elongation February 17, June 16, and October 9.

In addition to times of maximum elongation, there will be retrograde events. Mercury retrograde is probably one of the most potent planetary influences for investors and traders to be aware of. We often hear about Mercury retrograde events in mundane astrology. Classical astrologers will tell clients to not sign important contracts during Mercury retrograde, to not cross the street, to not leave their houses and so on. While I tend to ignore this mundane talk, I have noticed a striking correlation between financial market behavior and Mercury retrograde events.

In addition to times of maximum elongation, there will be retrograde events. Mercury retrograde is probably one of the most potent planetary influences for investors and traders to be aware of. We often hear about Mercury retrograde events in mundane astrology. Classical astrologers will tell clients to not sign important contracts during Mercury retrograde, to not cross the street, to not leave their houses and so on. While I tend to ignore this mundane talk, I have noticed a striking correlation between financial market behavior and Mercury retrograde events.

To understand the science of Mercury retrograde, consider the diagram in Figure 18.

In 30 days of time, planet Earth (shown as the larger circles in the diagram) will travel 30 degrees of the zodiac (from point 1 to point 2). But, Mercury is a faster mover. In the same 30 days of time, Mercury (shown as the smaller circles) will travel through about 120 degrees of the zodiac (point 1 to point 2), passing by Earth in the process. From our vantage point on Earth, as Mercury is setting up to pass Earth, we see Mercury in the sign of Aries. These sign determinations are made by extending a line from planet Earth through Mercury to the outer edge of the zodiac wheel. As Mercury completes its trip past Earth, we see it in the sign of Capricorn. In other words, the way we see it here on Earth, Mercury has moved backwards in zodiac sign as it passed Earth.

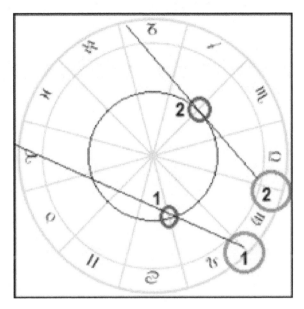

Figure 18
The concept of Mercury retrograde

Sometimes Mercury retrograde events encompass a sharp market inflection point. Sometimes a market peak or bottom will follow closely behind a retrograde event; sometimes a peak or bottom will immediately precede a retrograde event. Mercury retrograde events can be highly unpredictable.

The power players that influence the financial markets can use retrograde events at their whim to turn the tables in their favor. Knowing that a Mercury retrograde event is approaching, individual traders and investors must remain alert for trend changes. A chart technical indicator such as the Wilder Volatility Stop is helpful in determining if the price trend is changing.

Figure 19 illustrates 2020 and 2021 Mercury retrograde events overlaid on a chart of the E-mini S&P 500. The magnitude of some of these retrograde drawdowns was likely suppressed by the liquidity being injected into financial systems by central bankers around the world. As we enter 2022, expect to see tapering become the focal point of the central bankers. This may serve to heighten the magnitude of retrograde impacts on equity markets.

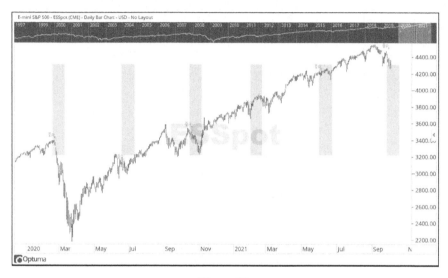

Figure 19
E-mini S&P 500 and Mercury retrograde

Mercury retrograde events can often be also seen influencing commodity markets. Figure 20 illustrates the correlation between Copper futures prices and Mercury retrograde. While the average investor may not be aggressively trading Copper futures, this correlation could be used to better manage price risk of copper-related mining stocks in an investment portfolio. Looking at the following chart of Copper futures, one can see the price low in early 2021 gave way to a rally under the influence of Mercury retrograde. A price drawdown that started in May, 2021 ended during the retrograde event in June, 2021.

Figure 21 illustrates Gold futures prices overlaid with Mercury retrograde events in 2021. Gold started 2021 on a strong note. Price then declined and entered a sideways consolidation pattern. A retrograde event, however, started a waterfall event that saw Gold price retreat $200 per ounce. In June 2021, Gold's attempt to rally strongly was thwarted by a retrograde event. Price weakness in September 2021 was arrested by the onset of the 3rd retrograde event of 2021.

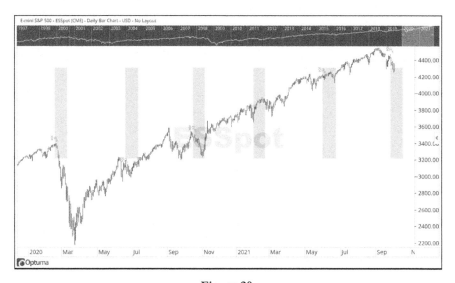

Figure 20
Mercury retrograde and Copper futures

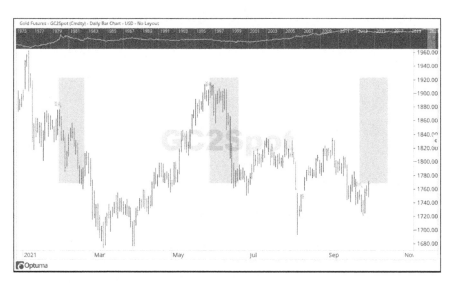

Figure 21
Mercury retrograde and Gold futures

For 2022, Mercury will be:

- ☿ retrograde from January 14 through February 3
- ☿ retrograde from May 10 through June 2
- ☿ retrograde from September 9 through October 2.

CHAPTER SEVEN
Professor Weston's Cycles

In 1921 a mysterious person from Washington, D.C. going by the name of Professor Weston wrote a paper in which he analyzed copious amounts of past price data from the Dow Jones Average. He applied cosine Fourier mathematics to the data and came up with a general set of rules for anticipating price action on the Dow Jones Average.

Who exactly Weston was, I will likely never know; another one of those figures who emerged to write his ideas down before vanishing into the ether.

Perhaps Weston knew W.D. Gann personally. Perhaps he just knew of him. In any case, Weston followed the 20-year Gann Master Cycle cycle of Jupiter and Saturn.

Weston further broke this long cycle into two components of 10 years.

He believed investors can expect:

✿ a 20-month market cycle to begin in November of the 1ˢᵗ year of the 10-year cycle

✿ another 20-month cycle to begin in November of the 5ᵗʰ year of the 10-year cycle

✿ 28-month cycles to begin in July of the 3ʳᵈ and 7ᵗʰ years of the 10-year cycle

✿ a 10-month cycle to begin in November of the 9ᵗʰ year of the 10- year cycle

✿ a 14-month cycle to begin in September of the 10ᵗʰ year of the 10-year cycle.

To put all this into perspective, the new Master Cycle began on October 30, 2020 as heliocentric Jupiter and Saturn made a 0-degree aspect. The entire cycle will run until November 2040.

Following Weston's methodology:

✿ the first 20-month cycle will start in November 2020 and go to until July 2022

✿ a 28-month cycle will run from July 2022 through November 2024

✿ a 20-month cycle will then follow until July 2026

✿ a 28-month cycle will run July 2026 through November 2028

✿ a 10-month cycle will then run through until September 2029

✿ lastly, a 14-month cycle will last until November 2030.

Weston also postulated that in the various years of a 10-year segment of the overall Master Cycle, there would be market maxima as listed in Figure 22.

YEAR OF CYCLE	MAXIMA	MAXIMA
1	March	October
2		May
3	January	September
4	April	November
5	May	November
6		June
7	January	September
8		June
9	April	
10	February	August

Figure 22
Weston's Secondary Cycles

Weston calculated these events using cycles of Venus. He argued that the 16th Harmonic of a 10-year period (120 months) was actually the heliocentric time it takes for Venus to orbit the Sun (120 x 30 / 16 = 225 days).

After the new Cycle begins in November 2020, Weston's work cautions investors to be alert for market maxima in March 2021 and in October 2021. The March 2021 period may have had the makings for a market maxima event but the liquidity from central bankers reigned supreme to hold any drawdown to 3%. The October 2021 maxima event was likewise held in check and limited to a 5% drawdown. May of 2022 should see another maxima event. May 2022 will align with the Federal Reserve ending its monthly liquidity injections. Will the projected maxima align to a market taper-tantrum?

Looking back to 2020 which was the end year in the previous Master Cycle, Weston's predictions call for a maxima in February, 2020 and another in August, 2020. The equity markets peaked in February, 2020 as fears over COVID-19 began to manifest. Massive fiat money stimulation then drove the markets from their March lows towards an over-valued maximum in late August. Starting September 1 2020, the S&P 500 pulled off over 300 points over the ensuing 3 weeks.

The markets truly are an overlapping array of cycles, just as Weston postulated back in the 1920s. The question that remains is, do these cycles just unfold in and of themselves? Or, do manipulative players working behind the curtain play a role?

With the aid of a spreadsheet program, you can use Weston's cosine model for yourself. To start with:

- ✿ Take a look at a multi-year chart of the stock or commodity you are interested in.

- ✿ Identify two major price tops separated by many years. Call this timespan 'A' years.

- ✿ Within the A years of time, identify the various price pullbacks that were equal to or greater than 10% in magnitude. Let the number of such drawdowns be 'N'.

Using Gold as an example, I note a top in 1980 and another in 2011 for a span of 31 years. A = 31. I count 6 drawdowns of 10% or more in this timespan. N=6.

N/A = 31/6 = 5.16 years.

I further note that Gold seems to have displayed a cycle of high to low from 1980 to 2000 which is 20 years.

In Weston's method, 20/5.16 = 3.87. In round figures, Weston would call the 3.87 value simply 4.00. Weston would then say that the 20-year timespan is in round figures the 4th harmonic of 5.16 years.

Weston would then calculate 4 x 5.16 = 20.64 years = 247.68 months.

The 4th part of 247.68 is 61.92.

360 / 247.68 = 1.45 degrees.

360/61.92 = 5.81 degrees.

In a spreadsheet, compute the cosine value at every 1.45 degrees and extend your compilations for 100 or more lines. Be careful to watch how your spreadsheet wants the data. You may have to convert the degrees to radians.

In another column, compute the cosine value at every 5.81 degrees and extend your compilations for 100 or more lines.

Next create a column of numeric values that extend from 1 to over 100. In an adjoining column, sum the two previously calculated cosine values.

You will now have the basis for an x-y plot using the spreadsheet software.

Study the curve that has been generated. Note the time intervals (months) from highs to lows. From a high price point on the curve of Gold, count forward the computed number of months and you will have a good idea of when to expect a low.

To capture a copy of Weston's actual paper, please see the following link:

https://bradleysiderograph.com/premium/members/1-new-techniques/2018-08-01-professor-weston-manuscript-locked-in-w-d-ganns-safe/

The following images in Figures 23 and 24 illustrate the spreadsheet and the resulting x-y plot.

	First Cosine Curve				2nd Cosine Curve				Sum	
x	cos	degrees	radians	x	cos	degrees	radians			
1	0.99968	1.45	0.025307	1	0.994863	5.81	0.101404	1	1.994543
2	0.998719	2.9	0.050615	2	0.979505	11.62	0.202807	2	1.978224
3	0.997119	4.35	0.075922	3	0.954084	17.43	0.304211	3	1.951203
4	0.994881	5.8	0.101229	4	0.91886	23.24	0.405615	4	1.913741
5	0.992005	7.25	0.126536	5	0.874196	29.05	0.507018	5	1.866201
6	0.988494	8.7	0.151844	6	0.820551	34.86	0.608422	6	1.809045
7	0.98435	10.15	0.177151	7	0.758476	40.67	0.709825	7	1.742825
8	0.979575	11.6	0.202458	8	0.688608	46.48	0.811229	8	1.668183
9	0.974173	13.05	0.227765	9	0.611665	52.29	0.912633	9	1.585839
10	0.968148	14.5	0.253073	10	0.528438	58.1	1.014036	10	1.496586
11	0.961502	15.95	0.27838	11	0.439782	63.91	1.11544	11	1.401284
12	0.95424	17.4	0.303687	12	0.346608	69.72	1.216844	12	1.300849

Figure 23
Cosine Spreadsheet

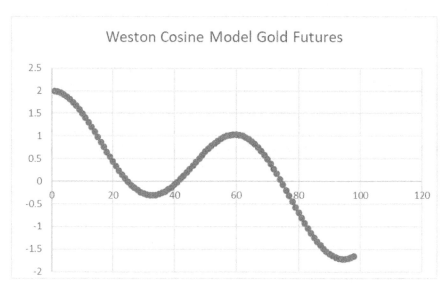

Figure 24
Cosine Plot

CHAPTER EIGHT

Shemitah and Kaballah Cycles

Classical astrologers measure time by synodic cycles and sidereal cycles.

A sidereal cycle is one that is measured from the vantage point of the Sun. If an observer were standing on the Sun, he or she would see Venus travel around the Sun one complete time in 225 days. Mars would take 687 days. The outer planets would take much longer. In fact, Saturn would take 29.42 years, Uranus 83.75 years, Neptune 163.74 years, and Pluto 245.33 years.

A synodic cycle is measured from the vantage point of here on Earth. To an observer standing here on terra firma, the time it takes Venus to record a conjunction with Sun until that same conjunction occurs again is 584 days. Mars takes 780 days from Sun/Mars conjunction to the next Sun/Mars conjunction. Saturn takes 376 days and the other outer planets 367 to 399 days.

The sidereal cycles of the outer planets bear an alignment to larger events in history. For example, the year 1776 is key to American history.

Add a Neptune cycle to 1776 and one gets to 1939 when the world was on the cusp of World War II. The American Civil War started in 1861 with the events at Fort Sumter. Add a Uranus cycle to this date and one gets to the time when World War II ended. Add another Uranus cycle and that takes us to 2028. Are we headed for another major conflict?

More intriguing than synodic and sidereal cycles are those cycles rooted in religious doctrine. Cycles related to the world of religion have a curious way of intersecting with financial market turning points. One religious concept is that of Shemitah which is rooted in the Hebrew Bible. I first learned of the Shemitah from the writings of Rabbi Jonathan Cahn.[1][2][3] On the surface, Cahn appears to be an average ordinary Rabbi from New Jersey, USA. But behind the scenes, he has done a masterful job of applying Shemitah to the financial markets. His books include: The Harbinger, The Book of Mysteries, and The Paradigm.

As Cahn explains, in the book of Exodus (Chapter 23, verses 10-11), it is written: You may plant your land for six years and gather its crops. But during the seventh year, you must leave it alone and withdraw from it.

In the book of Leviticus (Chapter 25, verses 20-22), it is written: And if ye shall say: "What shall we eat the seventh year? Behold, we may not sow, nor gather in our increase"; then I will command My blessing upon you in the sixth year, and it shall bring forth produce for the three years. And ye shall sow the eighth year, and eat of the produce, the old store; until the ninth year, until the produce come in, ye shall eat the old store.

Breaking these Biblical statements down into simple-to-understand terms means that every 7th year something will happen on the financial markets.

The first Shemitah year in the modern State of Israel was 1951-52 Subsequent Shemitah years have been 1958–59, 1965–66, 1972–73, 1979–80, 1986–87, 1993–94, 2000–01, 2007–08, and 2014-15. We are now in a Shemitah year which will continue through August of 2022.

A Shemitah Year starts in the month of Tishrei (the first month of the Jewish civil calendar) and ends in the month of Elul. The Gregorian calendar equivalent will have the 2021 Shemitah year starting on the evening before September 6.

The chart in Figure 25 illustrates the S&P 500 with some recent Shemitah years overlaid. Those who understand Shemitah could have profited handsomely from these moves on the S&P 500.

Figure 25
E-mini S&P 500 and Shemitah years

Shemitah years are not always about the equity markets. The chart in Figure 26 illustrates Oil prices with Shemitah years overlaid. Those who understood and used Shemitah as an investing timing strategy made serious money on the Crude Oil market in 2007 and again in 2014.

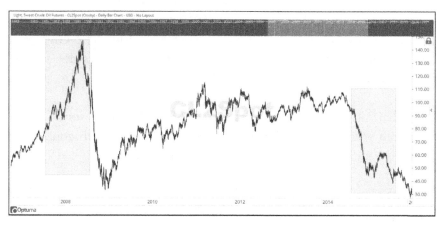

Figure 26
Crude Oil and Shemitah years

The key to using Shemitah is to quickly understand what the theme of the Shemitah year will be. Since the start of Shemitah in September 2021, it appears that the theme(s) might be rising interest rates and rising inflation/commodity prices.

Referring back to the Exodus and Leviticus Biblical passages, the message is there shall be no crop in the Shemitah year. In the year after the Shemitah, people shall live on the bounty of the crop produced in year 6, the year immediately prior to Shemitah. In the 9th year, the newly planted crop will be abundant.

The suggestion for here and now is, live off the bounty reaped up to September of 2021, for there will not be a bountiful crop for the next two years.

Massive fiat money stimulus drove the US equity markets to ridiculous levels in 2021 following the initial COVID panic in 2020. A bounty indeed for those institutions who availed themselves of the easy Fed QE money.

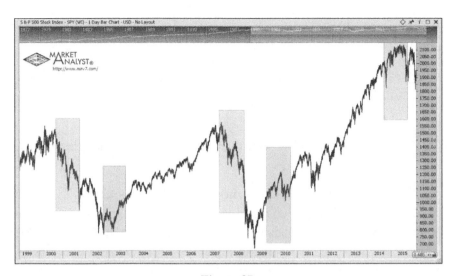

Figure 27
The Aftermath of Shemitah years

In Figure 27, which illustrates price action on the S&P 500, note the gap following the 2007-2008 Shemitah event. This gap is followed by another rectangle to mark what effectively is the second year after Shemitah. The key point is in the year immediately after Shemitah the market can fall and take out the price lows of the Shemitah year. The earlier-cited Biblical passages advise that we are to be living off the produce harvested (financial gains made) prior to Shemitah. Once the post-Shemitah period has elapsed, Figure 27 shows that the market can regain its footing and start to advance again.

The period September 2021 through September 2022 will be a Shemitah year. The post-Shemitah year will conclude by September 2023. Structure your trading and investing activity accordingly. Strictly adhering to the Shemitah cycles, profits ideally should have been taken by the September 2021 period. Plan to re-enter the markets in a serious way in September 2023. Along the way, look for only short term trading of oversold opportunities across various segments of the market.

According to Rabbi Cahn, in addition to the Shemitah cycle, there are certain dates from the Hebrew calendar that can have a strong propensity to align with swing highs and lows on the New York Stock

Exchange. Rabbi Cahn advises to pay close attention to four particular dates from the Hebrew calendar:

☼ The 1st Day of the month of Tishrei marks the start of the Jewish civil calendar, much like January 1 marks the start of the Gregorian calendar.

☼ The 1st day of the month of Nissan marks the start of the Jewish sacred year.

☼ The 3rd important date is the 9th day of the month of Av which marks the date when Babylon destroyed the Temple at Jerusalem in 586 BC. Other calamitous events have beset the Jewish people on the 9th of Av throughout history. In particular, Cahn tells of the mass expulsion of Jewish people from Spain in 1492. As this expulsion was going on, a certain explorer with 3 ships was about to set sail on a voyage of discovery. That explorer sailed out of port on August 3, 1492 which was one day after the 9th of Av. That explorer was Christopher Columbus; he found the New World and as Cahn tells it, the New World became a new home for the Jewish people.

☼ The 4th key date in the Jewish calendar is Shemini Atzeret (The Gathering of the Eighth Day). This date typically falls somewhere in late September through late October in the month of Tishrei.

The website www.chabad.org allows one to quickly scan back over a number of years to pick off these important dates. I have examined price action on the Dow Jones Average across several years in the context of these key dates. My backtesting has shown a variable correlation to these dates. But the correlation is valid enough that I recommend traders and investors pay attention to these dates.

For 2020, these four key dates fell as follows: 1st of Nissan on March 26, the 9th of Av on July 30, the 1st of Tishrei on September 19 and Shemini Atzeret on October 10.

The 1st of Nissan aligned perfectly to the March 2020 market sell-off lows. Not much happened at the 9th of Av. The 1st of Tishrei marked a swing low on the S&P 500. Shemini Atzeret marked a swing high on the S&P 500. These correlations are further evidence that these dates are critical to watch.

For 2021, these dates fell as follows: 1st of Nissan on March 14, the 9th of Av on July 18, the 1st of Tishrei on September 7, and Shemini Atzeret on September 28.

The 1st of Nissan and the 9th of Av aligned to minor pullbacks in the context of a rising market fueled by central bank liquidity. The 1st of Tishrei, however, was a different story. This date marked a peak in the S&P 500 Index. The Shemini Atzeret date was a stiff down-day and reinforced that the trend had definitely turned negative.

For 2022, these critical dates will be: 1st of Nissan on March 26, the 9th of Av on July 30, the 1st of Tishrei on September 19 and Shemini Atzeret on October 10.

In addition to cycles related to religious doctrine, cycles related to Kaballah mathematics are equally as intriguing. A 2005 article in Trader's World magazine suggested that W.D. Gann might have been instilled with knowledge of Jewish mysticism based on the Kabbalah. Gann apparently had connections to a New York personality called Sepharial who is said to have taught Gann about astrology and esoteric matters.

The Kabbalah centers around the Hebrew Alef-bet (alphabet). The Hebrew Alef-bet comprises 22 letters. In Kabbalistic methodology, these letters are assigned a numerical value. Starting with the first letter, values are 1, 2 3 4 5 6 7 8 9 10 20 30 40 50 60 70 80 90 100 200 300 and 400.

There are many mathematical techniques that can be applied to parsing the Alef-bet. One in particular involves taking the odd-numbered letters and the even numbered letters and assigning their appropriate numerical values.

The numerical value (sum total) of the Alef-bet is 1495. The sum total of the odd-numbered letters is 625. The sum total of the even numbered letters is 870.

- ✪ 625/1495 = 42%. Taking a circle of 360 degrees, 42% is 150.5 degrees.

- ✪ 870/1495 = 58%. Taking a circle of 360 degrees, 58% is 209.5 degrees.

- ✪ Kabbalists are also well aware of phi as it pertains to the Golden Mean. Phi is famously known as 1.618.

- ✪ 1/phi = 62%. Taking a circle of 360 degrees, 62% is 222.5 degrees.

- ✪ 1 – (1/phi) = 48%. Taking a circle of 360 degrees, 48% is 137.5 degrees.

From a significant price low (or high), one can examine price charts for time intervals when a geocentric or heliocentric planet advances 137.5, 150.5, 209.5 or 222.5 degree amounts. My back-testing has shown that heliocentric advances are more apt to align to market turning points.

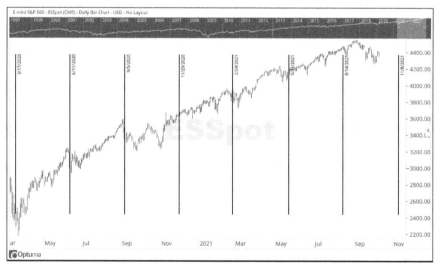

Figure 28
Kabbalistic Math and Venus 137.5 degrees

To illustrate, consider the significant low in March 2009 on US equity markets (S&P 500) as a start point. Using the phi approach, Figure 28 shows that from the March 2009 lows, 11 years later, 137.5-degree advancements of heliocentric Venus land spot on the March 2020 lows. Figure 28 also shows a market high in early September 2020 aligned to a Venus 137.5-degree interval.

In 2021, 137.5-degree heliocentric Venus intervals came into play on the S&P 500 on February 24, May 21, August 14, and November 8.

Extending 209.5-degree intervals forward from the March 2009 lows reveals that one of these intervals lands right at the September 2021 start of Shemitah.

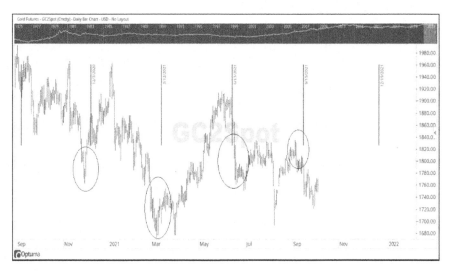

Figure 29
Kabbalistic Math and Gold

To further illustrate the effect of these intervals, consider the 2011 significant high for Gold. Over nine years later, Figure 29 shows how the 150.5-degree advancements of heliocentric Venus continue to align themselves with uncanny accuracy to various short term trend swings on Gold price.

For 2022, watch for 150.5-degree Venus intervals to land at: March 16,

June 19, September 20 and January 30, 2023.

To add to the mystery, applying 137.5-degree advancement intervals from the 2011 high shows a perfect alignment to the September 2020 interim top in Gold price.

For 2022, watch for Venus 137.5-degree intervals to possibly impact Gold price trend on January 6, March 31, June 26, and September 19.

As a final example, from the 2016 multi-year low on Wheat futures, the application of the 222.5-degree intervals of Venus yields a very compelling alignment to swing highs and lows. In 2021, these 222.5-degree intervals landed within days of a March swing low and within days of an August swing high.

For 2022, watch the time around June 4, and September 1 for possible impacts to the price trend of Wheat futures.

Once you source your heliocentric planetary data, either from an on-line site or from a purchased ephemeris book, you can look at practically any commodity or market index for a significant high or low. Apply the 137.5, 150.5, 209.5, or 222.5 degree intervals of Venus forward in time from that start point. You should see a striking pattern emerge.

In addition to Venus, my back-testing has also shown that movements of heliocentric Mars from a past high or low event also have significant potential to align to market turning points. Although not shown here, extending heliocentric intervals of 209.5 degrees forward from the March 2009 lows produces various intervals, one of which lands precisely at the February 2020 highs that faltered as the Covid pandemic started.

CHAPTER NINE

Lunar Declination, Pythagoras, and Musical Harmony

The Moon's orbit around the Earth can be described in terms of synodic and sidereal cycles. The sidereal period of the Moon is 27.5 days and the synodic period is 29.53 days. This latter period lends itself to the expression lunar month.

During each sidereal lunar cycle, the Moon can be seen to vary in its position above and below the lunar ecliptic. In a sidereal period, the Moon will go from maximum declination to maximum declination. It is said that W.D. Gann was a proponent of following lunar cycles when trading Soybeans and Cotton. The question is, was he doing so based on the premise that the Moon's gravitational pull governs the ocean tides? With our bodies being substantially water, was he postulating that the Moon was influencing the emotions of Soybean and Cotton traders? Or, was he part of a larger contingent of traders who were timing the markets to their own advantage at Moon declination highs and lows?

Figure 30 presents a chart of Soybeans with the Moon declination in the lower panel. Note the propensity for declination maxima and minima to align to trend changes. A 4-hour chart would be the ideal way to study these price swings closely.

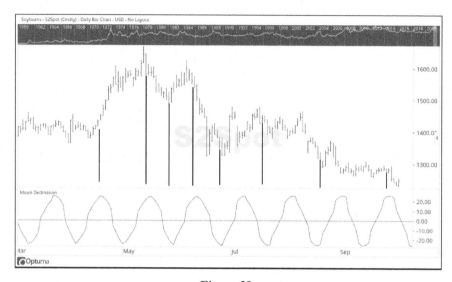

Figure 30
Moon Declination and Soybeans

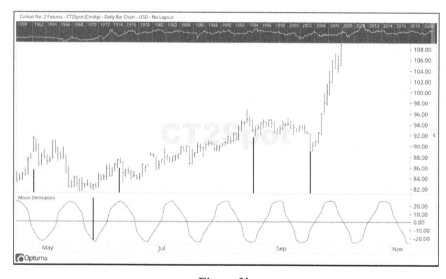

Figure 31
Moon Declination and Cotton

Figure 31 illustrates Cotton futures prices during 2021 to date. I have overlaid the chart with a dark vertical line at some maximum, minimum and zero-point lunar declination events. You can now see why a trader like W.D. Gann would have used lunar declination phenomenon to give himself an advantage in the Cotton and Soybean markets.

If you are a trader of commodity futures, I encourage you to study price charts of your favorite futures contracts to see if a Moon declination correlation exists. If it does, take advantage of it.

To assist you in some backtesting of your own, consider:

In 2020, Moon recorded maximums in its declination on January 10, February 6, March 5, April 1, April 29, May 25, June 22, July 19, August 16, September 12, October 9, November 5, December 4, and December 30.

In 2020, declination minima occurred January 23, February 19, March 18, April 14, May 11, June 7, July 5, August 1, August 29, September 25, October 22, November 18, and December 16.

In 2021, Moon was at its maximum declination: January 1, January 27, February 23, March 22, April 19, May 16, June 12, July 9, August 6, September 2, September 29, October 27, November 23, December 20.

In 2021, Moon was at minimum declination: January 13, February 9, March 8, April 4, May 1, May 29, June 26, July 23, August 20, September 15, October 12, November 9, and December 6.

For 2022, Moon will be at its maximum declination: January 17, February 13, March 12, April 8, May 6, June 2, June 29, July 26, August 23, September 19, October 15, November 13, December 10.

For 2022, Moon will be at its minimum declination: January 30, February 26, March 25, April 22, May 19, June 15, July 13, August 9, September 6, October 3, October 30, November 26, December 24.

There are also cyclical intervals that relate to Pythagorean mathematics. Recall from high school math class that the Pythagorean Theorem says for a right angled triangle, $c^2 = a^2 + b^2$. Also, recall the peculiar feature 'pi' which carries the value 1.34159.

The following mathematical explanations use this theorem and pi to identify cyclical intervals.

Imagine a square with each side being 1000 units long. Now imagine a circle that is inscribed into that square. The diameter of that circle will be 1000 units.

The circumference of the circle inscribed into the square will be pi x diameter. In this case pi x 1000 = 3141.159 units.

The perimeter of the square will be the sum total of the four sides, 1000 + 1000 + 1000 +1000 = 4000.

The ratio of the perimeter to the circumference of the circle is 4000 ÷ 3.14159 = 1273.4.

Next, imagine the square cut diagonally across. If the sides of the square are each 1000 units, the length of the diagonal is defined by the Pythagorean equation $a^2 + b^2 = c^2$. The diagonal is thus defined by $(1000)^2 + (1000)^2 = c^2 = 1414.21$ units. The ratio of the diagonal to the perimeter of the square is 1414.21/4000 = 0.3536.

Next imagine a square with sides equal to 1273. The diagonal calculates as $(1273)^2 + (1273)^2 = c^2 = 1800.3$ units.

Now multiply 0.3536 by 1800.3 to get 636.6.

Imagine a right angled triangle with one side being 1 unit long and another side being 2 units long. The Pythagorean expression $(1)^2 + (2)^2 = c^2$ can be used to define the length of the longest side (the hypotenuse). In this case, c = 2.2361 units. Now, add the length of all three sides and divide by the length of the longest side (the hypotenuse). This works out to $(1 + 2 + 2.2361) \div 2.2361 = 2.3416$ units.

Consider next the sum of the two sides divided by the hypotenuse. This works out to $(2 + 1) \div 2.2361 = 1.3416$.

A circle contains 360 degrees. Another unit of measure for expressing this is radians where 360 degrees equals 2 x pi radians. In our right angled triangle with sides of 1, 2, and 2.2361 units, the angles in that triangle will necessarily be 30, 60, and 90 degrees. An angle of 60 degrees equals (pi $\div$ 3) radians.

Consider a circle of diameter 1000 units inscribed in a square with sides of 1000 units. The expression (pi x 1000) $\div$ 3 = 1047.2 units.

Double this figure and one gets 2094.4 units.

Consider a right angled triangle with sides of 1047.2 and 2094.4. The Pythagorean theorem says the length of the hypotenuse is $(1047.2)^2 + (2094.4)^2 = c^2$. Solving for c yields 2341.6.

Taking all of the above calculated numbers in these various complicated mathematical expressions and rounding them off slightly gives us the following sequence of numbers:

637, 1047, 1273, 1341, 1414, 1800, 2094, 2341, and 4242.

I suggest that anyone interested in financial astrology take the time to read W.D. Gann's 1927 book entitled *Tunnel Through the Air*. [1] This book is part sci-fi fantasy, part romance story and part war story. Peppered throughout are curious references to times when the main figure Robert Gordon executed trades on Cotton, Wheat and a stock called Major Motors. In one part of the story, Gann draws from the biblical book of Daniel and lists a couple cyclical intervals that Robert Gordon used in trading.

The book of Daniel says: And from the time that the daily sacrifice is taken away, and the abomination of desolation is set up, there shall be one thousand two hundred and ninety days.

The book of Daniel also says: Blessed is he who waits, and comes to the one thousand three hundred and thirty-five days. Looking at the above list of Pythagorean related numbers, the figure 1273 is close to the 1290 days referenced in Daniel. The figure 1341 is close to the 1335 referenced in Daniel.

Taking the above list of numbers to be trading days (not calendar days), from a significant low (or high) on a commodity, a stock or an index, one can plot these intervals on the chart. One will often notice a very high propensity for the intervals to align to points of trend change.

To illustrate, consider the price chart of Gold in Figure 32 where I have selected the December 2015 price low at $1050 per oz as a start point. Note how the various intervals align to trend change inflection points on the chart.

Figure 32
Gold and Ermanometry Intervals

If you are wondering where this bit of complex math comes from, credit goes to Bill Erman who passed away in 2016. He called his mathematical approach Ermanometry. As another example, consider Soybeans and their 2016 price high. Figure 33 presents some of the Ermanometry intervals. Note the curious alignment to points of trend change. The 1414 period interval aligns to a trend reversal on Beans price in 2020. The 1800 interval aligns to the second part of what was a double top.

Figure 33
Soybeans and Ermanometry Intervals

Turning to the US equity markets, for the Dow Jones Average, if one starts the interval count at the December 2018 lows, the 637 interval lands mere days ahead of the October 2020 corrective pullback. The 1047 interval will land November 24, 2021.

For the Nasdaq 100, if one starts the interval count at the early 2016 corrective lows, the 1047 interval lands right at the late 2018 corrective pullback. The 2094 interval will land in early December, 2021 as Figure 34 shows. Such are the hidden mysteries of geometry.

Study a price chart of interest to you. Identify a significant high or low turning point and then apply the calendar day intervals 637, 1047, 1273, 1341, 1414, 1800, 2094, 2341, and 4242 days. You now have a powerful market timing tool at your fingertips.

Figure 34
E-mini Nasdaq and Ermanometry Intervals

In my 2021 travels, I found an interesting book at a used book store. The title is *Sacred Geometry* by author Robert Lawlor. [2]

Lawlor reminds us that when people are asked to compute a mean between two numbers (a and b) they instinctively use the method (a + b)÷2. This is how the North American educational school system regards the concept of a mean. This method is called the arithmetic mean.

Lawlor goes on to describe the geometric mean. Take 2 extreme numbers (a and c). The geometric mean equates as: b2 = ac.

Lastly, he describes the harmonic mean which equates as: b = (2ac)÷(a + c).

Lawlor briefly touches on the notion of combining arithmetic means and harmonic means. Consider the two extreme values of 1 and ½. The value of 1 is double that value of ½. An interval where one extreme is double that of the other extreme is termed an octave in musical terms. The arithmetic mean of 1 and ½ is ¾. The harmonic mean of 1 and ½ equates to 2/3. Combining these numbers gives 1, ¾, 2/3, ½. Taking the inverse of these numbers, one can create a rising sequence of 1, 4/3, 3/2, 2. Lawlor notes that in musical harmony an octave would be what the ear interprets if a guitar string of length 1 and a string twice as long are plucked. Mid-way in this octave, 4/3 is called the perfect 4th and 3/2 is called the fifth.

Let's now focus on these means and musical notes in the context of the financial markets.

Consider the stock Beyond Meat (N:BYND) as an example. Figure 35 has been overlaid with a series of horizontal lines. The start point is the March 2020 lows at $48. The musical progression of 1, 4/3, 3/2, 2 yields the following: $48, $64, $72, $96. As the chart shows, price action following the Covid panic sell-off quickly rebounded through these levels. After $96, a new octave is started and the numerical progression will be: $96, 127.96, $144, $192.

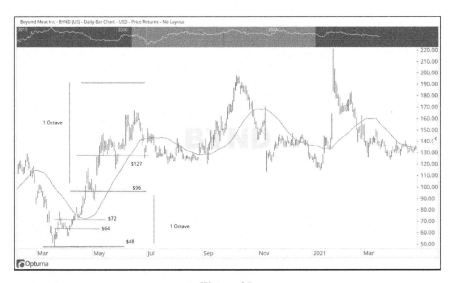

Figure 35
N:BYND and musical harmony

But…there is a problem. Price action does not make it to the $127 level. The harmony has failed. This is a warning sign for a short-term trader to perhaps initiate a stop loss order.

The price high at the day when price action fails to push higher was $116.64. If price is now to trend down, price action will follow the musical progression of 1, ¾, 2/3, ½. In other words, $116.64, $87.48, $77.76, $58.30. As Figure 36 shows, harmony enters the picture and price falls to $88.51 which is just shy of the harmonic $87.48.

Figure 36
N:BYND and musical harmony

From the actual low of $88.51, price action hints that it might want to move higher. If it does move higher, it should ideally unfold according to harmony: 1, 4/3, 3/2, 2. In price format this should turn out to be: $88.51, $118, $132,76, $177. But, as mid-June, 2020 enters the picture, it is obvious that the octave harmony has been disrupted. The disrupting factor is Mercury turning retrograde.

From a high of $167.16, it looks as though price action might want to ease off. If it does, it ideally should follow the progression of 1, ¾, 2/3, ½, or $167.16, $125.27, $111.44, $83.58. In Figure 37, note what happened. Price action fell and found support in harmony with Nature at the $125 level.

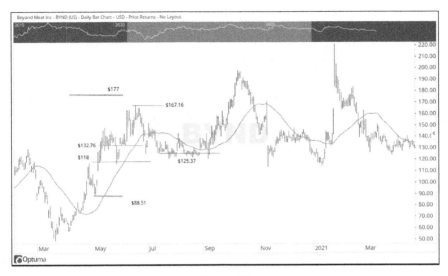

Figure 37
N:BYND and musical harmony

Figure 38
N:BYND and musical harmony

As I work to complete this manuscript in October 2021, the price of BYND is trending down and has been doing so since early July 2021. Again, Mercury retrograde seems to have been the catalyst for disrupting harmony. From the June 30, 2021 high of $157.20, harmony should see price action hitting $117.90, $104.80 and $78.60. As the chart in Figure 38 shows, price action is currently trading in harmony with Nature at the $104 level. From here, either price action gives a buy signal (crosses above 34- day average, or the Wilder indicator produces a buy signal), or price declines to complete the harmonic at $78.60.

CHAPTER TEN

Planetary Declination

As the various celestial bodies make their respective journeys around the Sun, they can be seen to move above and below the celestial equator plane. Celestial bodies experience declinations of up to about 25 degrees above and below the celestial equator plane. Declinations above the equator plane are positive. Declinations below the equator plane are negative.

Declination can be viewed one planet at a time or by pairs of planets. Suppose that at a particular time Mars is at 10 degrees of declination above the celestial equator and at that same time Venus is at 9 degrees of declination. Factor in an allowance for up to 1.5 degrees tolerance in the measurement of declinations. It could then be said these two planets are at parallel declination.

Suppose that at a given time Jupiter is at 5 degrees of declination above the celestial equator. At that same time Pluto is at 6.5 degrees declination below the celestial equator. Again, allow for up to 1.5 degrees of tolerance. It could then be said that Jupiter and Pluto are at contra-parallel declination. Parallel and contra-parallel events can have a powerful bearing on the financial markets.

I have spent many hours exploring parallel and contra-parallel events with respect to both individual stocks and commodity futures contracts. This research was spurred on by a 25-year-old astrology book I found in a used bookshop in early 2017. [1] In studying this old book, I have found there are some commodities that bear a correlation to the same parallel and contra-parallel conditions that were in place at the date the contracts first started trading on an exchange (first trade date).

To use this strategy, obtain the planetary declinations at the first trade date for the stock or commodity in question. Examine the data for evidence of parallel and contra parallel events. Watch for these parallel and contra-parallel occurrences to repeat themselves at future dates. To illustrate, consider the following Gold, Cotton, Crude Oil, and Soybeans studies:

Gold

Gold futures started trading in New York on Dec 31, 1974. At that date, the planets were at the following declinations relative to the ecliptic:

Sun	-23.06 degrees
Venus	-22.42 degrees
Mars	-22.45 degrees
Jupiter	-7.39 degrees
Saturn	+22.05 degrees
Uranus	-11.37 degrees
Neptune	-20.3 degrees
Pluto	+11.44 degrees

Looking closer at these numbers reveals

- ✿ Uranus is contra-parallel to Pluto (contra-parallel= declinations within 1.5 degrees of each other, but signs are opposite)

- ✿ Saturn is contra-parallel Mars

- ✿ Saturn is contra-parallel Venus

- ✿ Saturn is contra-parallel Sun

- ✿ Mars is parallel Sun (parallel = declinations with 1.5 degrees of each other, signs are the same)

- ✿ Mars is parallel Venus

- ✿ Sun is parallel Venus.

For 2021:

- ✿ Saturn was contra-parallel Mars from February 1 to 23 and June 26 to July 13

- ✿ Saturn was contra-parallel Venus from April 27 to May 8 and June 14 to July 13

- ✿ Saturn was contra-parallel Sun from March 4 to May 14

- ✿ Mars was parallel Sun from May 29 to June 10 and September 18 to October 14

- ✿ Mars was parallel Venus May 18 to 30 and July 5 to July 26

- ✿ Sun was parallel Venus for 4 days either side of June 4

- ✿ Saturn was contra-parallel Venus January 1 to 12, March 21 to April 22, and June 9 to 30.

The price chart of Gold in Figure 39 has been overlaid with some of these events from 2021. Note the correlation between market swing points and these declination events.

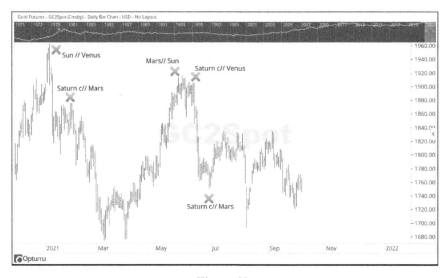

Figure 39
Gold and Declination Phenomenon

In *Tunnel Through the Air*,[2] Gann references days when the story hero Robert Gordon was very certain that a trend change would occur on his various Major Motors trades. My instinct was to check planetary declination at Major Motors first trade date and compare those figures to Robert Gordon's key dates. Just as I suspected, there is a declination connection.

Gann hints strongly that instead of watching for two planets to become parallel or contra-parallel so as to match the situation at the first trade date, one should instead be watching for Mars and Venus in particular to simply pass the same declination level as they were at on the first trade date.

In the case of Gold, the first week of January 2021 should have delivered a trend change response as Venus passed -22.42 degrees declination. As Figure 38 shows, this indeed did happen. December 3-19, 2021 and December 18-31, 2021 should also deliver price trend responses.

For 2022, Mars and Venus will be at the same declination they were at in December 1974:

○ January 1 through March 1 (Mars) and Venus November 20 to December 31.

Cotton

Cotton futures started trading in New York on June 20, 1870. At that date, the planets were at the following declinations relative to the ecliptic:

Sun	+23.45 degrees
Venus	+14.84 degrees
Mars	+21.42 degrees
Jupiter	+21.32 degrees
Saturn	-22.09 degrees
Uranus	+22.31 degrees
Neptune	+6.86 degrees
Pluto	+2.99 degrees

Taking a cue from W.D. Gann and *Tunnel Through the Air*, in 2021 Mars was at its 1870 first trade declination from February 20 to March 15 and again from May 28 to June 21. Venus was at its 1870 declination from April 20-28 and from July 15-22.

The price chart of Cotton in Figure 40 has been overlaid with these events from 2021. The events in the first half of the year align to trend swing points. The Venus event in July adds to the tenor of the existing trend. The price explosion in September at first glance is not related to Mars or Venus. A deeper examination reveals that Mercury was retrograde as this upsurge got underway. The fascinating thing is, Mercury was also retrograde in September 1870. W.D. Gann's approach to declination definitely has merit.

For 2022, Mars and Venus will be at the same declination they were at in June 1870:

☿ June 7 to 15 (Venus) and Mars August 29 to October 14 (Mars).

Figure 40
Cotton and Declination Phenomenon

Crude Oil

Crude Oil futures started trading in America on March 30, 1983. At that date, the planets were at the following declinations relative to the ecliptic:

Sun	+3.52 degrees
Venus	+16.28 degrees
Mercury	+4.38
Mars	+9.52 degrees
Jupiter	-21.18 degrees
Saturn	-9.87 degrees
Uranus	-21.70 degrees
Neptune	-22.20 degrees
Pluto	+5.42 degrees

For 2021, Venus was at the +16.28 degrees of declination level from April 24 to May 2 and July 8 to 19. Mars was at +9.52 degrees of declination August 5 to 17.

The price chart of Crude Oil in Figure 41 has been overlaid with these events.

Following the *Tunnel Through the Air* methodology, Crude made trend changes in close alignment with Mars and Venus declinations.

For 2022, Mars and Venus will be at the same declination they were at in June 1983:

☿ June 12 to 21 and August 22-30 (Venus) and Jun 29-July 10 (Mars).

Figure 41
Crude Oil and Declination Phenomenon

Soybeans

As will be discussed in a coming chapter, the Chicago Board of Trade was founded April 3, 1848. W.D. Gann was very cognizant of this date when trading Soybeans. Looking at Soybean futures through the lens of this date as opposed to the 1936 date when Soybean futures actually started trading yields some interesting finds. At the 1848 date, the planets were at the following declinations relative to the ecliptic:

Sun	+5.35 degrees
Venus	-7.26 degrees
Mercury	-6.10
Mars	+24.80 degrees
Jupiter	+23.26 degrees
Saturn	-6.08 degrees
Uranus	+6.48 degrees
Neptune	-11.48 degrees
Pluto	-5.46 degrees

For 2021, Venus was at the -7.26 degrees of declination level from March 6 to 12, and August 29-September 4. Mars was at the +24.80 level March 17 to May 24. It is intriguing to note that the significant trend change in early May 2021 aligns to the Mars declination.

Figure 42
Soybeans (1848) and Declination Phenomenon

In October 1936 when the Soybeans contract first started trading in Chicago, Mars was at +10.32 degrees of declination. It is curious to note that in early August 2021, Mars was passing through this declination level. Mars seems to have cut short an effort at a recovery rally.

The price chart of Soybeans in Figure 42 has been overlaid with these various events.

For 2022:

○ Venus will be at the -7.26 degree level of declination from April 12 to 21.

○ Mars will be at the +24.80 degree level of declination from October 14 through year end.

The New York Stock Exchange (NYSE) can also be viewed through the lens of declination. As the next chapter will reveal, the NYSE traces its origins to May 17, 1792. At that date Venus was at +11.42 degrees declination and Mars was at +5.42 degrees of declination. It is intriguing to learn that in late February, 2020 full-blown panic was setting in across equity markets. Venus was approaching its historical declination of +11.42 degrees declination.

W.D. Gann was very generous to leave us clues in *Tunnel Through the Air* as to how he used planetary declination. Historical declination data is readily available on the internet. Or, a software program such as *Solar Fire Gold* will also help you identify the data.

CHAPTER ELEVEN
NYSE 2022 Astrology

No examination of the astrology of the New York Stock Exchange would be complete without mention of Louise McWhirter. After years of reading old papers and manuscripts, I still have no idea who Louise McWhirter was. What I do know, is during her lifetime, Louise McWhirter focused intently on the astrology of the New York Exchange. Her technique which revolves around the New Moon (lunation) remains viable to this day. The one caveat I must offer is that since the Covid 19 panic selloff, the US Federal Reserve has been pumping massive amounts of liquidity into the financial system. This liquidity has partly obscured weakness on equity markets with 'buy the dip' being the resounding cry when markets do attempt to express weakness. As the Federal Reserve starts to taper back its liquidity, I hope that 2022 will allow McWhirter's method to shine through once again.

The Lunation and the New York Stock Exchange

A lunation is the astrological term for a New Moon. At a lunation, the Sun and Moon are separated by 0 degrees which means the Sun and Moon are together in the same sign of the zodiac. The correlation between the monthly lunation event and New York Stock Exchange price movements was first popularized in 1937 by McWhirter. In her book, Theory of Stock Market Forecasting,[1] she discussed how a lunation making hard aspects to planets such as Mars, Jupiter, Saturn and Uranus was indicative of a coming month of volatility on the New York Stock Exchange. She also paid close attention to Mars and Neptune, the two planets that rule the New York Stock Exchange. The concept of planetary rulership extends back into the 1800s with each zodiac sign having a celestial body that rules that sign. McWhirter said those times of a lunar month when the transiting Moon makes 0 degree aspects to Mars and Neptune should be watched carefully. McWhirter arrived at her Mars and Neptune rulership conclusion by observing that the 10th House of the 1792 birth horoscope wheel for the NYSE spans Pisces and Aries. Neptune rules Pisces and Mars rules Aries.

New York Stock Exchange – First Trade Chart

The New York Stock Exchange officially opened for business on May 17, 1792. As the horoscope in Figure 43 shows, the NYSE has its Ascendant (Asc) at 14 degrees Cancer and its Mid-Heaven (MC) at 24 Pisces.

Figure 43
NYSE First Trade horoscope

McWhirter further paid close attention to those times in the monthly lunar cycle when the transiting Moon passed by the NYSE natal Asc and MC locations at 14 Cancer and 24 Pisces respectively.

Horoscope Charts and the McWhirter Methodology

In my research and writing, I follow the McWhirter methodology for shorter term trend changes. When forecasting whether or not a coming month will be volatile or not for the NYSE, the McWhirter methodology starts with creating a horoscope chart for the New Moon date and positioning the Ascendant of the chart at 14 degrees Cancer (the Ascendant position on the 1792 natal chart of the New York Stock Exchange). Positioning the Ascendant is made easy in the Solar Fire Gold software program. Aspects to the lunation are then studied. If the lunation is at a 0, 90, or 120 degree aspect to Mars, Neptune, 14 Cancer, or 24 Pisces, one can expect a volatile month ahead. A lack of such aspects portends a less volatile period. The McWhirter method demands a consideration of where the Moon is at each day. Aspects of the Moon to Mars, Neptune, 14 Cancer, or 24 Pisces represent dates of potential short term trend reversals. Although not expressly stated by McWhirter, it is also important to pay attention to those dates when Moon is at either maximum or minimum declination. As well, dates when Mercury is retrograde and dates when Venus is at or near its maximum or minimum declination should be considered carefully.

Similarly, when studying an individual stock or an individual commodity futures contract, the McWhirter approach calls for the creation of a horoscope chart at the first trade date of the stock or commodity. The Ascendant is then shifted so that the Sun is at the Ascendant. The software program Solar Fire Gold is very good for generating first trade horoscope charts for McWhirter analysis where the Ascendant needs to be shifted.

In stock and commodity analyses, McWhirter paid strict attention to those times of a calendar year when transiting Sun, Mars, Jupiter, Saturn, Neptune, and Uranus made hard 0, 90, and 180 degree aspects to the natal Mid-Heaven, natal Ascendant, natal Sun, natal Jupiter and even the natal Moon of the individual stock or commodity future being studied.

One must be alert at these aspects for the possibility of a trend change, the possibility of increased volatility within a trend, or even the possibility of a breakout from a chart consolidation pattern. Evidence of such trend changes will be found by watching price action relative to moving averages and by utilizing oscillator type functions (MAC-D, DMI, RSI and so on).

McWhirter Lunation February-March 2020 – a Powerful Example

The McWhirter method can best thoroughly appreciated by examining events leading up to the 2020 Covid panic sell-off. A New Moon on February 23, 2020 came during Mercury retrograde. Within a day of the New Moon, the Moon transited past the NYSE natal Mid Heaven (24 Pisces) and also past co-ruler Neptune. A day later, Moon transited past co-ruler Mars. This was a lot of planetary energy compressed into a short timeframe. The NYSE reacted in a strong negative manner. An attempt to counter the negative sentiment failed as Moon made a hard 90-degree aspect to Neptune and a 0-degree aspect to 14 of Cancer on March 2 and 5 respectively. On March 24, a New Moon event again arrived. This lunation was within orb of being conjunct to NYSE co-ruler Neptune. With that bit of energy, the markets began a sharp recovery. As a trader or investor, you can view the events of February-March, 2020 through the lens of the Federal Reserve and fiat stimulus or you can view the events through the lens of the New Moons and aspects to key NYSE natal zodiac points.

What follows in this chapter is a listing of the date for each lunar cycle in 2022 along with a list of times when Moon passes Mars, Neptune, the NYSE natal Mid-Heaven at 24 Pisces, and the NYSE natal Ascendant at 14 Cancer.

Since the last edition of the Almanac, I have become acquainted with one additional lunar feature that has a curious alignment to the financial markets. The feature is called Moon Void of Course. Moon is considered Void of Course (VOC) from the time it makes no aspects to other planets to the time it enters the next zodiac sign. In a given month, the Moon will be VOC perhaps 12 times or so. In order for VOC to affect equity markets, the VOC event must occur Monday through Friday, must be more than 4 hours in duration, and must occur during NYSE trading hours. I disregard any VOC events outside these parameters. The net result is that in a typical month there might be up to four VOC events. In the following discussion of lunation events, I have including VOC events.

2022 Lunation Events

January 2022

Market action in January 2022 will be influenced by the New Moon cycle that commences on January 2, 2022 with Sun at 12 Capricorn. NYSE co-ruler Mars is positioned a benign 30 degrees away. The lunation is in alignment with the NYSE 1792 natal Descendant and is 180 degrees opposite the natal Ascendant. This cycle is highlighted by having Mercury and Venus in retrograde which can lead to trend volatility. Venus will record its Inferior Conjunction early in the month. In addition, January will see the North Node change signs and enter Taurus. This lunar cycle runs until January 31, 2022 and could be greatly energized.

Key dates during this lunar cycle are:

- ✿ January 5: Moon VOC

- ✿ January 7: Moon passes Neptune and natal mid-Heaven point

- ✿ January 7: Mercury at greatest easterly elongation

- ✿ January 8: Venus at Inferior Conjunction. It will appear as the Morning Star later in January

- ✿ January 10: Moon VOC

- ✿ January 14: Mercury turns retrograde

- ✿ January 16: Moon passes 14 of Cancer. This is a Sunday. Watch for a market reaction the Friday before or on Monday January 17

- ✿ January 17: Moon at maximum declination

- ✿ January 26: Mars at its minimum declination

- ✿ January 28: Venus retrograde event is complete

- ✿ January 29: Moon passes NYSE co-ruler Mars. This is a weekend. Watch for a market reaction immediately before or after the weekend

- ✿ January 30: Moon at minimum declination.

February 2021

The February New Moon cycle commences on February 1, 2022 with Sun at 12 Aquarius. This lunar cycle runs until March 1, 2022. There are no detrimental aspects to key points visible.

Key dates during this lunar cycle are:

- ✿ February 1: Moon VOC

- ✿ February 3: Mercury retrograde ends

- ✿ February 4: Moon passes co-ruler Neptune

- ✿ February 11: Moon VOC

- ✿ February 13: Moon passes 14 of Cancer and makes its maximum declination

- ✿ February 16: Moon VOC

- ✿ February 17: Mercury at its greatest westerly elongation

- ✿ February 25: Moon VOC

- ✿ February 26: Moon makes minimum declination

- ✿ February 27: Moon passes NYSE co-ruler Mars

- ✿ March 1: Noon VOC.

March 2021

The March New Moon cycle commences on March 2, 2022 with Sun at 12 Pisces. This lunar cycle runs until March 31, 2022.

Key dates during this lunar cycle are:

- ☼ March 3: Moon passes co-ruler Neptune and natal mid-Heaven point of 24 Pisces

- ☼ March 8: Moon VOC

- ☼ March 11: late in the day, Moon passes 14 of Cancer

- ☼ March 12: Moon at declination maximum

- ☼ March 15: Moon VOC

- ☼ March 24: Moon VOC

- ☼ March 25: Moon at declination minimum

- ☼ March 28: early in the morning, Moon passes co-ruler Mars

- ☼ March 28: Moon VOC

- ☼ March 31: Moon passes co-ruler Neptune and natal mid-Heaven point of 24 Pisces.

April 2021

The April New Moon cycle commences on April 1, 2022 with Sun at 11 Aries. This lunar cycle runs until April 29, 2022. The lunation itself is within orb of being a hard 90 degrees square to 14 of Cancer (natal NYSE Ascendant). This hard aspect could energize this cycle.

Key dates during this lunar cycle are:

- ☼ April 8: Moon at maximum declination.

- ☼ April 8: Moon VOC

- ☼ April 9: Moon passes 14 of Cancer.

✿ April 22: Moon at declination minimum.

✿ April 26: Moon passes NYSE co-ruler Mars.

✿ April 29: Mercury at greatest easterly elongation.

May 2021

The May New Moon cycle commences on April 30, 2022 with Sun at 10 Taurus. This lunar cycle runs until May 29, 2022. The lunation itself is not in any hard aspect to other planets. Mars is a favorable 60 degrees to the lunation.

Key dates during this lunar cycle are:

✿ May 5: Venus passes through 0 degrees of declination

✿ May 6: Moon passes 14 of Cancer and hits maximum declination

✿ May 9: Moon VOC

✿ May 10: Mercury turns retrograde

✿ May 19: Moon at minimum declination

✿ May 24: Moon passes co-ruler Neptune and natal Mid Heaven point of 24 Pisces. Moon also passes co-ruler Mars.

June 2022

The June New Moon cycle commences on May 30, 2022 with Sun at 9 Gemini. This lunar cycle runs until June 28, 2022. The lunation itself is not in harmful aspect to any other planets. Watch the timeframe around July 21 in particular.

Key dates during this lunar cycle are:

✿ May 30: Mars passes through 0 degrees declination

✿ June 2: Mercury retrograde ends

- ✿ June 2: Moon passes 14 of Cancer and makes maximum declination

- ✿ June 14: Moon VOC

- ✿ June 15: Moon at minimum declination

- ✿ June 16: Mercury at greatest westerly elongation

- ✿ June 20: Moon passes co-ruler Neptune and natal Mid Heaven point of 24 Pisces

- ✿ June 22: Moon passes co-ruler Mars.

July 2022

The July New Moon cycle commences on June 29, 2022 with Sun at 7 of Cancer. This lunar cycle runs until July 27, 2022. The lunation itself is arguably within orb of being conjunct to the 14 of Cancer point.

Key dates during this lunar cycle are:

- ✿ June 29: early in the morning, Moon passes 14 of Cancer and makes a declination maximum

- ✿ July 13: Moon at minimum declination

- ✿ July 18: Moon passes co-ruler Neptune and the natal Mid Heaven point

- ✿ July 20: Moon VOC

- ✿ July 21: Moon passes co-ruler Mars

- ✿ July 22: Venus at its declination maximum. Any effect on the markets will be noted about 10 days either side of the exact maximum

- ✿ July 25: Moon VOC

- ✿ July 26: Moon at maximum declination.

August 2021

The August New Moon cycle commences on July 28, 2021 with Sun at 5 Leo. This lunar cycle runs until August 26, 2022. The lunation itself is not at a disagreeable aspect to other planets.

Key dates during this lunar cycle are:

- ✡ August 8: Mon VOC
- ✡ August 9: Moon at minimum declination
- ✡ August 12: Moon VOC
- ✡ August 14: Moon passes co-ruler Neptune and natal 24 Pisces point
- ✡ August 19: Moon passes co-ruler Mars
- ✡ August 23: Moon passes 14 Cancer and makes declination maximum.

September 2022

The September New Moon cycle commences on August 27, 2022 with Sun at 4 Virgo. This lunar cycle runs until September 24, 2022. The lunation is at a 90 degree square to NYSE co-ruler Mars. This month will mark the end of the Shemitah Year as outlined in an earlier chapter. This month will be further highlighted by Mercury at greatest westerly elongation, Mercury retrograde and Venus at Superior Conjunction. This could be a highly energized month.

Key dates during this lunar cycle are:

- ✡ August 27: Mercury at greatest easterly elongation
- ✡ August 31: Moon VOC
- ✡ September 6: Moon at declination minimum
- ✡ September 8: Moon VOC

- ✿ September 10: Moon passes Neptune and natal Mid Heaven
- ✿ September 10: Mercury turns retrograde
- ✿ September 15: Moon VOC
- ✿ September 16: Moon passes co-ruler Mars
- ✿ September 19: Moon passes 14 of Cancer and makes declination maximum
- ✿ September 22: Moon VOC.

October 2022

The October New Moon cycle commences on September 25, 2022 with Sun at 3 Libra. This lunar cycle runs until October 24, 2022. This month starts off with Mercury retrograde ending.

Key dates during this lunar cycle are:

- ✿ October 3: Moon at declination minimum
- ✿ October 8: Moon passes NYSE co-ruler Neptune and natal Mid Heaven
- ✿ October 9: Mercury at greatest westerly elongation
- ✿ October 10: Moon VOC
- ✿ October 15: Moon at declination maximum and Moon passes NYSE co-ruler Mars
- ✿ October 16: Moon passes 14 of Cancer
- ✿ October 20: Moon VOC
- ✿ October 22: Venus at Superior Conjunction. After a Superior event, Venus will appear as the Evening Star.

November 2022

The November New Moon cycle commences on October 25, 2022 with Sun at 2 Scorpio. This lunar cycle runs until November 22, 2022.

Key dates during this lunar cycle are:

- ✿ October 30: Mars turns retrograde
- ✿ October 30: Moon at declination minimum
- ✿ November 4: Moon passes Neptune and natal Mid Heaven
- ✿ November 11: Moon passes co-ruler Mars
- ✿ November 13: Moon passes 14 Cancer and makes declination maximum
- ✿ November 21: Moon VOC.

December 2022

The December New Moon cycle commences on November 23, 2022 with Sun at 1 Sagittarius. This lunar cycle runs until December 22, 2022.

Key dates during this lunar cycle are:

- ✿ November 26: Moon at declination minimum
- ✿ December 2: Moon passes Neptune and natal Mid Heaven
- ✿ December 8: Moon passes NYSE co-ruler Mars
- ✿ December 8: Mars at declination maximum. The effect on the market might be seen 7-10 days either side of this exact maximum
- ✿ December 10: Moon passes the 14 of Cancer point and completes its declination high
- ✿ December 13: Moon VOC
- ✿ December 14: Venus at declination minimum. The effect on the market might be seen 7-10 days either side of this exact minimum
- ✿ December 21: Mercury at its greatest easterly elongation
- ✿ December 23: Moon nearing declination minimum.

January 2023

The January New Moon cycle commences on December 23, 2022 with Sun at 3 Capricorn. Shortly after this cycle begins, Mercury will turn retrograde and a new calendar year will begin. The calendar year 2023 could be interesting as post-Shemitah years can be troublesome, as explained earlier in this Almanac.

Reminder - Follow the Trend

At the risk of sounding overly repetitive, I must stress again the importance of following the trend. A question that I routinely get is when during one of these astrological transit events should a person implement a trade? The answer is very simple. You should consider implementing a trade when you see the trend change. Always let the trend be your friend. I am sure you have heard this mantra before.

There are many ways of measuring trend. My experience has shown me that the methodologies developed by J. Welles Wilder are very powerful for identifying trend changes. In particular I prefer to use his Wilder Volatility Stop. Wilder's 1978 book, New Concepts in Technical Trading Systems, is a highly recommended read if you are seeking to learn more about his methods. Another technique for gauging trend is to overlay a price chart with moving averages, such as the 34-day and 55-day ones. These are Fibonacci numbers and I routinely use them in my personal examination of trend.

CHAPTER TWELVE

Commodities 2022 Astrology

Gold

Investors who own Gold or related mining stocks are accustomed to routinely checking the price of Gold by tuning into a television business channel or perhaps obtaining a live on-line quote of the Gold futures price. What many do not realize is quietly working behind the scenes to define the price of Gold is an archaic methodology called the *London Gold Fix*.

The 1919 Gold Fix

The London Gold Fix occurs at 10:30 am and 3:00 pm local time each business day in London. Participants in the daily fixes are: Barclay's, HSBC, Scotia Mocatta (a division of Scotia Bank of Canada) and Societe Generale. These twice daily collaborations (some would say collusions) provide a benchmark price that is then used around the globe to settle and mark-to-market all the various Gold-related derivative contracts in existence.

The history of the Gold Fix is a fascinating one. On the 12th of September 1919, the Bank of England made arrangements with N.M. Rothschild & Sons for the formation of a Gold market in which there would be one official price for Gold quoted on any one day. At 11:00 am, the first Gold fixing took place, with the five principal gold bullion traders and refiners of the day present. These traders and refiners were N.M. Rothschild & Sons, Mocatta & Goldsmid, Pixley & Abell, Samuel Montagu & Co. and Sharps Wilkins.

Figure 44
1919 London Gold Fix horoscope

The horoscope in Figure 44 depicts planetary positions at this date in history. Observations that jump off the page include: North Node had just changed signs, Venus was retrograde, Sun and Venus were conjunct, Mercury and Saturn were conjunct, Mars, Neptune and Jupiter were all conjunct at/near the Mid-Heaven point of the horoscope, and Saturn was 180 degrees opposite Uranus.

Gold investors who have been around for a while will remember the significant $800 per ounce price peak recorded by Gold in January 1980. To illustrate how astrology is linked to Gold prices, consider that at this price peak the transiting North Node had just changed signs and was 90 degrees hard aspect to the natal Node in the 1919 horoscope. Consider too that Mars and Jupiter were both coming into a 0-degree conjunction with the natal Sun location in the 1919 horoscope.

For those who were involved in Gold more recently, recall that Gold hit a significant peak in early September 2011 at just over $1900/ounce. At that peak, Sun and Venus were conjunct to one another as they were in the 1919 Gold Fix horoscope. What's more, they were within a few degrees of being conjunct to the natal Sun location in the 1919 horoscope.

In the few weeks that followed this 2011 peak, Gold prices plunged nearly $400 per ounce. But, then Gold found its legs again and began to rally. This rally seems directly related to Mars coming into a 0-degree conjunction to the Mars-Jupiter-Neptune location of the 1919 horoscope wheel.

In early August, 2020 Gold again made a significant high, getting just above $2000/ounce. Sun conjunct to the 1919 Mars position along with Mars conjunct to the 1919 natal Moon are the obvious factors at work.

Such is the complex nature of Gold prices. I have studied past charts of Gold and I am amazed at how many price inflection points are related in one way or another to the astrology of the 1919 Gold Fix horoscope wheel. To those readers who are of the opinion that Gold price is manipulated, your notion is indeed a valid one. I believe that astrology is the secret language being spoken amongst those that play a hand in the price manipulation.

For 2022, January will see the Node changing signs at a time when it is 180 degrees opposite the 1919 natal Node. The month of May will see Venus passing the 1919 natal Moon location. June will have Mars doing likewise. August will see Venus passing the 1919 natal Jupiter, natal Neptune and natal Mars points. September will see Venus pass the 1919 natal Sun location. Mars will be retrograde from October onwards and will not be making 1919 aspects.

Declination

Declination further adds to the intrigue. At September 12, 1919 Mars was at -1.5 degrees, Venus was also at -1.5 degrees, Moon was at 0 degrees declination and Sun was at +4 degrees declination. If four celestial bodies all at or near 0 degrees declination seems like a co-incidence, it is not. This date in 1912 was picked for these obvious astrology features.

Examination of a Gold price chart for 2021 shows that these declination levels do align to short term trend changes, especially when viewed in pairs (ie Moon at 0 declination and Venus at or near 0 declination at the same time).

☼ For 2022, Moon will be at 0 declination: January 9, January 23, February 5, February 19, March 4, March 18, April 1, April 14, April 28, May 12, May 25, June 9, June 22, July 6, July 19, August 2, August 15, August 29, September 11, September 25, October 9, October 23, November 5, November 20, December 3 and December 17.

☼ For 2022, Mars will be at or near 0 degrees declination around June 1. Venus will be at or near 0 degrees declination around May 5 and again October 1.

☼ For 2022, Sun will be at or near 4 degrees declination around March 30 and again around September 12.

1974 Futures

Gold futures contracts started trading in America on the New York Mercantile Exchange on December 31, 1974. Figure 45 illustrates the planetary positions in 1974 at the first trade date of Gold futures.

Note that in the 1919 chart Mars and Neptune are conjunct one another. In the 1974 chart, Mars and Neptune are also conjunct one another.

Figure 45
Gold futures First Trade horoscope

Next, consider why the New York Mercantile Exchange would launch a new futures contract on December 31, a time when most staff would be off for Christmas holidays. If this seems more than a bit odd, you are not alone in your thinking.

Note the location of Moon in the 1974 horoscope at 12 degrees of Leo. Now, look at the 1919 horoscope and observe that 12 degrees of Leo is where Mars and Neptune are located. The 1919 Gold Fix Ascendant is located 90 degrees square to the 1974 Futures Ascendant if one assumes a first trade start time of 9:00 am which is a more than reasonable assumption. I take these curious placements as further evidence of a deliberately timed astrological connection between Gold price, the 1919 Gold Fix date, and the 1974 first trade date for Gold futures. All very intriguing stuff to be sure.

Venus and Mars transiting the 1974 natal horoscope

Times when Venus and Mars transit past key points in the 1974 first trade horoscope deserve attention.

In early January 2021, Venus passing natal Sun aligned to a Gold price pullback. In early March 2021, Venus passing the 1974 natal Jupiter aligned to a price reversal low. In mid-June, 2021, Mars moving past the 1974 natal Moon location aligned to a $150/ounce drawdown on Gold prices. In mid-June as well, Venus passing the 1974 natal Saturn contributed to the efforts of Mars.

✿ In early January 2022, Mars will pass conjunct to the 1974 natal Mars location.

✿ In early February 2022, transiting Mars will pass the 1974 natal Sun location.

✿ As Venus emerges from retrograde in late January, it will find itself conjunct the 1974 natal Sun.

✿ In mid-April, Venus will pass the 1974 natal Jupiter point.

✿ In late July 2022, Venus will pass the 1974 natal Saturn point.

✿ In mid-August, Venus will pass the natal Moon point.

Mercury Retrograde

Another valuable tool for Gold traders to consider is Mercury retrograde events. Watch for technical chart trend indicators to suggest a short term trend change at a retrograde event.

The chart in Figure 46 illustrates the connection between these Mercury phenomena and Gold prices in 2021. Not every Mercury retrograde event will result in a dramatic price swing. I remain of the opinion that the power players on Wall Street and in London reserve the right to use Mercury retrograde events to their advantage as they see fit. Traders and investors must remain alert for possible trend changes at Mercury retrograde events.

For 2022, Mercury will be:

✿ retrograde from January 14 through February 3

✿ retrograde from May 10 through June 2

✪ retrograde from September 9 through October 2.

Figure 46
Mercury retrograde and Gold prices

The declination intrigue extends to the 1974 date. At December 31, 1974 Mars and Venus were at -22 degrees declination and Sun was at -23 degrees declination. Surely not a co-incidence to select a first trade date with three celestial bodies all at the same declination.

✪ For 2022, Mars will be at or near its minimum declination in late January.

✪ Venus will be at or near -23 degrees declination in mid-December.

Silver

Silver futures started trading on a recognized financial exchange in July 1933. Figure 47 shows the First Trade horoscope for Silver futures in geocentric format.

Figure 47
Silver futures First Trade horoscope

I am intrigued with this first trade date. I am sure Silver could have started trading anytime in 1933. July 4, 1776 is a critical date in US history and on this date Sun was at 14 Cancer. Recall that 14 Cancer also figures prominently in the first trade horoscope of the NYSE. In 1933, markets would have been closed for the 4th of July celebrations. A first trade date of July 5, is as close as authorities could come to the July 4 date. On July 5, Sun at 13 degrees is within a degree of the critical 14 of Cancer point.

My research has shown that times when transiting Sun, transiting Mars and transiting Jupiter make hard aspects to the natal Sun point at 12 degrees Cancer should be watched carefully for evidence of trend changes and price inflection points.

Figure 48 illustrates some 2021 aspects to 14 Cancer.

☼ For 2022, Mars will be 180 degrees to 14 Cancer in early February and 90 degrees to 14 of Cancer in mid-June.

☼ Sun will be 180 degrees to 14 Cancer in early January, at 90 degrees in early April, conjunct in July, and 90 degrees in October.

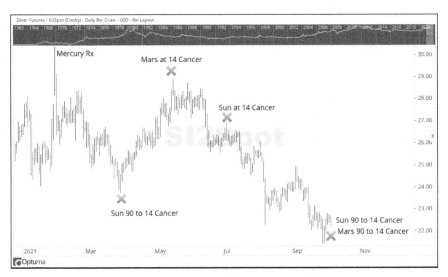

Figure 48
Silver futures and 14 of Cancer

Jupiter and the 1933 natal Sun

In April 2011, Silver prices reached a peak at just under $50 per ounce. Transiting Jupiter was making a 90-degree aspect to the 1933 natal Sun at the time. From this peak, Silver prices declined towards a significant low in late 2015. Along the way, transiting Jupiter made a 0-degree conjunction to the 1933 natal Sun in the August 2013 timeframe. Silver prices behaved extremely erratically during this period. During October, November and December 2016, Jupiter made a 90 degree hard aspect to the natal Sun position. Silver price moved quickly from $17 to $19 per ounce and then fell hard down to $15.50.

In early 2020, Silver prices again were impacted by Jupiter passing 180 degrees opposite to the 1933 natal Sun location. The end of this transit was marked by the start of a sharp decline in Silver prices, which just so happened to coincide with the COVID market sell-off. Was it COVID that caused Silver to sell off? Or was it market players lurking in the shadows manipulating markets in accordance with astrological cycles? Transiting Jupiter will again make a 90-degree aspect to the 1933 natal Sun in March 2023.

Declination

Planetary declinations should also be considered when studying price action of Silver futures. In particular, the declination maxima and minima of Venus and also of Sun should be watched. Venus had just made its maximum declination in 1933 as Silver futures were starting to trade for the very first time. Sun was at 22 degrees declination. Moon was at a declination minimum and Mars curiously enough happened to be at 0 degrees declination.

In 2021, Silver prices began to trend down in earnest following the early June Venus maximum declination event. Looking farther back in time, trend changes can be seen developing as Venus approached maximum declinations along with Moon at minimum declination. Times of Mars nearing 0 degrees declination with Moon at minimum declination showed a similar propensity for the start of trend changes.

- ✡ For 2022, Sun will be at its maximum declination at the Summer Solstice on June 21. Sun will at its minimum declination at the Winter Solstice on December 21.

- ✡ For 2022, Venus will exhibit its maximum declination either side of the July 23 date.

- ✡ For 2022, Mars will be at 0 degrees declination in late January.

Copper

The first trade date for Copper futures was July 29, 1988. Figure 49 illustrates the first trade horoscope in geocentric format.

Figure 49
Copper futures first trade horoscope

Inferior Conjunction

This horoscope wheel features an Inferior Conjunction of Mercury in the sign of Leo. Also, notice in this horoscope that the first trade date is that of a Full Moon.

An Inferior Conjunction of Mercury marks the start of a new Mercury cycle around the Sun. Mercury Inferior Conjunction events always occur in association with Mercury being retrograde.

Mercury Retrograde

Figure 50 shows Copper prices overlaid with Mercury retrograde events. Knowing that retrograde events are approaching, one should watch for short term trend changes using a suitable trend indicator.

Figure 50
Copper and Mercury retrograde events

For 2022, Mercury will be:

- ☿ retrograde from January 14 through February 3

- ☿ retrograde from May 10 through June 2

- ☿ retrograde from September 9 through October 2.

Copper prices are also influenced by Sun and Mars making aspects to the 1988 natal Sun and natal Mars points.

- ☿ For 2022, Sun will make 90-degree square aspects to natal Sun in late April and early November.

- ☿ Mars will make a 90-degree square to the 1988 natal Mars in mid-July.

Declination

At the 1988 first trade date, Mars was at -1.4 degrees declination. Mercury was at or near its maximum declination and Moon was at its minimum declination.

☿ For 2022, Mercury will be at or near maximum declination around May 5 and again around July 11.

Canadian Dollar, British Pound & Japanese Yen

These three futures instruments all started trading on May 16th, 1972 at the Chicago Mercantile Exchange. The horoscope in Figure 51 illustrates planetary placements at this date. It is interesting to note that Mars is 180 degrees opposite Jupiter. This suggests that Mars and Jupiter may play a role in price fluctuations on these currencies. Mars is also 0 degrees conjunct to Venus, suggesting another cyclical relationship.

Figure 51
Pound, Yen, Canadian First Trade horoscope

The Mars-Venus Influence

Mars conjunct Venus events only occur every couple years. A Mars/Venus conjunction event occurred in August 2019. For the latter half of 2019, the Canadian Dollar traded in a sideways pattern. The swing low recorded in August during the Mars/Venus conjunction proved to be the low (which was tested twice) for the entire latter half of 2019. A brief Mars/Venus conjunction will in July 2021 created a swing low.

In 2022, February and March will deliver a more extended Mars/Venus conjunction pattern. Watch these currencies closely.

Natal Mars and Natal Sun Transits

Transiting Sun passing natal Sun, natal Mars and quite often natal Jupiter are events that currency traders may wish to focus on.

To illustrate, the chart in Figure 52 illustrates the effect in 2021 on the British Pound of transiting Sun passing conjunct and square to natal Mars (2 degrees of Cancer in the 1972 horoscope chart) and Sun passing conjunct and square to natal Sun (25 of Taurus in the 1972 horoscope).

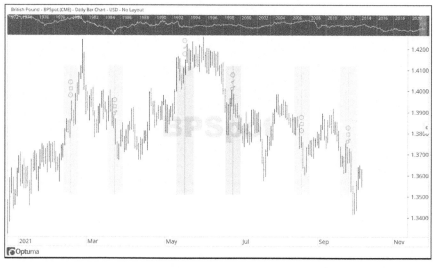

Figure 52
British Pound Sun passing natal Mars & natal Sun

The chart in Figure 53 illustrates the effect on the Canadian Dollar of transiting Sun and Mars making 0 and 90 degree aspects to the 1972 natal Sun and natal Mars.

For 2022, transiting Sun will make wide aspects to the natal Mars point of 2 degrees Cancer as follows:

✿ June 16 to June 29: 0 degrees conjunct

✿ March 16 to March 27: 90 degrees square

✿ September 18 to 30: 90 degrees square.

For 2022, transiting Sun will make hard aspects to the natal Sun point of 25 Taurus as follows:

✿ February 8 to 20: 90 degrees square

✿ May 8 to May 24: 0 degrees conjunct

✿ August 13 to August 24: 90 degrees square.

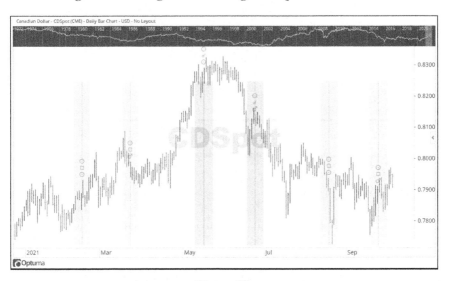

Figure 53
Canadian Dollar aspects to natal Sun and Mars

Mercury Retrograde

Currency traders should pay close attention to Mercury retrograde events as they can bear a good alignment to trend changes on the Pound, Yen and Canadian Dollar.

The Canadian Dollar price chart in Figure 54 has been overlaid with Mercury retrograde events. All too often there is a correlation to swing highs and lows.

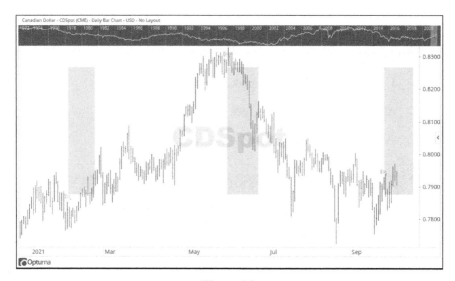

Figure 54
Canadian Dollar and Mercury retrograde

For 2022, Mercury will be:

- ✿ retrograde from January 14 through February 3
- ✿ retrograde from May 10 through June 2
- ✿ retrograde from September 9 through October 2.

Declination

At this 1972 first trade date, Mars, Venus and Moon were all at or near their declination maxima. A look at past price performance suggests Mars and/or Venus at or near declination maxima align to trend changes.

For 2022,

- ✿ Venus will exhibit its maximum declination either side of the

July 23 date

☼ Mars will be at or near its maxima in early December.

Euro Currency Futures

The Euro became the official currency for the European Union on January 1, 2002 when Euro bank notes became freely and widely circulated.

Figure 55
Euro Currency First Trade horoscope

Natal Transits

Events of transiting Sun making 0, and 90 degree aspects to the natal Sun position in the Euro 2002 First Trade horoscope are also worth watching as they often align to inflection points on the Euro. In 2021 notice how these transits all came within close proximity to price inflection points.

Figure 56
Natal Transits and the Euro Currency

For 2022, transiting Sun will be:

- ☿ At a 0 degree conjunction to natal Sun during the first week of January

- ☿ 90 degrees to natal Sun March 26 through April 7

- ☿ 90 degrees to natal Sun from 28 of September through October 9.

Declination

At this 2002 first trade date, Venus, Mercury and Sun were all at or near their declination minima. Moon was at its declination maxima. In early 2021, the trend on the Euro changed. Over the ensuing 3 months the Euro declined from the 1.23 level to the 1.17 level. The trigger point for this move lower was Venus and Mercury both being at their declination lows at the same time.

- ☿ For 2022, Venus will be at or near its declination low in mid-December.

☿ Mercury will be at or near its lows around the same time.

Watch closely for a significant trend change at these declination events.

Australian Dollar

Australian dollar futures started trading on the Chicago Mercantile Exchange on January 13, 1987. As the horoscope in Figure 57 shows, Sun and Mercury are at Superior Conjunction at 22-23 degrees Capricorn.

Figure 57
First trade horoscope of Australian Dollar futures

Mercury Retrograde

Times when Mercury is retrograde should be considered when trading Australian Dollar futures. The chart in Figure 58 has been overlaid with Mercury retrograde events. These events often align to price inflection points.

Figure 58
Australian Dollar and Mercury Conjunctions

For 2022, Mercury will be:

- ✧ retrograde from January 14 through February 3
- ✧ retrograde from May 10 through June 2
- ✧ retrograde from September 9 through October 2.

Declination

At this 1987 first trade date, Mercury was at or near its declination low. Moon was at its maximum. Mars was at or near 0 degrees of declination.

- ✧ For 2022, Mercury will be at or near its declination low in early December.

- ✧ •Mars will be at or near 0 declination around June 1.

30-Year Bond Futures

30-Year Bond futures started trading in Chicago on August 22, 1977. Figure 59 presents the geocentric first trade horoscope for this date.

Figure 59
First trade horoscope for 30-Year Bond futures

Natal Transits

Money and its cost (ie Bond yields) are critical factors to the functioning of an economy. I find it interesting that the 30 Year Bond natal horoscope has Mars at 24 Gemini. In the 1776 natal horoscope of the USA, Mars just so happens to be at 21 Gemini. Was this a factor in selecting the first trade date of the 30 Year Bond futures? Had the first trade date been set at the previous Friday (Aug 19), Mars would have been exactly at 21 of Gemini.

My research has indicated that events of transiting Mars making 0 and 90- degree aspects to the natal Mars position at 24 degrees Gemini are valuable tools for interpreting Bond prices. Also, Mars making aspects to natal Sun at 29 Leo is important to watch. Figure 60 illustrates Bond price performance with the Mars/natal Mars transits overlaid. Mars square natal Sun and Mars conjunct natal Mars have nicely bracketed the March Bond lows. A price high point was defined by Mars conjunct natal Sun in early August, 2021.

Figure 60
30-Year Bonds and Mars in aspect to natal Mars

For 2022, transiting Mars will make hard 0 and 90-degree aspects to natal Sun and natal Mars as follows:

- ☿ May 10 to 25, Mars will pass 90 degrees to natal Mars

- ☿ In October 2022, Mars will be 0 degrees to natal Mars

- ☿ August 10 to 28, transiting Mars Sun will make a 90 degree aspect to natal Sun.

Mercury Retrograde

In the first trade horoscope in Figure 59 note that the position of Mercury (at 20 Virgo) is further delineated by a letter S. This letter denotes stationary. This first trade date of August 22, 1977 comes one day prior to Mercury turning retrograde. Was this also a factor in selecting this first trade date? The price chart in Figure 61 has been overlaid with Mercury retrograde events. In early 2021, a brief attempt at a counter-trend rally was stopped by a retrograde event. A sudden price retreat in late May, 2021 was followed by a strong rally under the influence of retrograde.

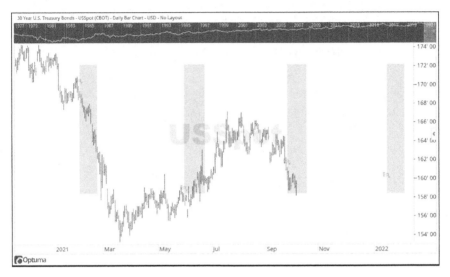

Figure 61
30-Year Bonds and Mercury retrograde

For 2022, Mercury will be:

☿ retrograde from January 14 through February 3

☿ retrograde from May 10 through June 2

☿ retrograde from September 9 through October 2.

Declination

At the 1977 first trade date, Mars was at its maximum declination and Mercury was at 0 degrees declination.

In early April, 2021 the 30 Year Bond futures began a move that took price from the 155 level to the 167 level. The trigger point for this move was Mercury at 0 degrees declination while Mars was closing in on maximum declination.

○ For 2022, Mars will reach maximum declination in early December.

○ Mercury will pass through 0 degrees declination April 1, August 22, September 27 and October 15.

10-Year Treasury Note Futures

10-Year Treasury Notes started trading in Chicago on May 3, 1982. Figure 62 presents the geocentric first trade horoscope for this date.

Figure 62
First trade horoscope for 10-Year Treasury Notes

Retrograde

Notice in the first trade horoscope in Figure 59, that Mars is denoted Rx which stands for retrograde. Therein rests another valuable clue. Looking back at past Treasury Note performance, significant price inflections at Mars retrograde events can be seen. Mars was retrograde from early September through to mid-November 2020. This event turned out to be a price high on the 10 Year Treasury market.

✿ In 2022, Mars will be retrograde from October 30 through early 2023.

Mercury retrograde events also bear watching when following price action on the 10-Year Treasury Notes. Figure 63 illustrates the connection between price inflection points and Mercury retrograde. In 2021, a retrograde event sparked a sizeable sell-off that lasted into April. The second wave of a push higher began at a retrograde event in June 2021.

Figure 63
Mercury retrograde and 10-Year Treasuries

For 2022, Mercury will be:

✿ retrograde from January 14 through February 3.

✿ retrograde from May 10 through June 2.

✿ retrograde from September 9 through October 2.

Declination

At the 1982 first trade date, Mars and Venus were within a degree of each being at 0 degrees declination. An examination of past price data reveals a solid alignment to these planets at 0 declination and short-term trend swings.

- ☿ For 2022, Mar will be at 0 degrees declination on June 1.

- ☿ Venus will be at 0 declination around May 6 and again around October 1.

Wheat, Corn, Oats

1877 Futures

Wheat, Corn and Oats futures all share the same first trade date from 1877. The horoscope in Figure 64 shows planetary placements at this date. Note how the 14 Cancer point appears exactly opposite the Ascendant point.

Figure 64
First trade horoscope for Wheat, Corn and Oats futures

CBOT 1848

W.D. Gann was also known to follow a first trade horoscope wheel from April 3, 1848, the date the Chicago Board of Trade was founded. Figure 65 shows this horoscope wheel.

In the 1877 Wheat/Corn natal horoscope, the Sun is at 12 Capricorn, exactly square to the location of Sun in the 1848 horoscope.

Figure 65
1848 first trade horoscope for CBOT

Natal Aspects

Events of transiting Sun making 0, 90 and 180-degree aspects to the natal Sun position in the 1877 first trade horoscope or the 1848 CBOT natal horoscope can be used as a tool to guide traders through the shorter term price volatility of Wheat, Corn and Oats. Because of the peculiar alignment of these two horoscopes, a 0 degree conjunction to natal Sun in the 1877 horoscope will be a 90 degree square to the natal Sun of the 1848 horoscope.

A more important tool is to consider 0 and 90 degree aspects of heliocentric transiting Jupiter to natal heliocentric Jupiter. In early 2016, a 90 degree event saw Corn prices rally almost $1 per bushel, which is $5000 on a single futures contract. A conjunct event in early 2019 saw a similar move. In early 2022, a 90-degree event will again occur.

The effect on Wheat prices is also apparent. In 2016, the 90 degree event saw a 50 cent per bushel drop in price at the transit. The conclusion of the 2019 transit saw an explosive $1.30 per bushel rally on Wheat prices.

Mars is also a planet to watch closely. In March 2020, Mars passing conjunct to the 1877 natal Sun location triggered a steep fall in Corn prices. In August 2020, a 90-degree square event sparked a rally in Corn prices which took Corn from just over $3 per bushel up to nearly $7.50 per bushel.

In 2021, transiting Mars was 90 degrees square to the 1877 natal Sun from September 28 to October 8.

In 2022,

- ✪ Mars will be conjunct the 1877 natal Sun from February 5 to 16

- ✪ Mars will be 90 degrees square the 1877 natal Sun from June 3 to 17.

Declination

Planetary declination also is a powerful tool that can be used to help traders identify the coming of trend reversals. A sizeable price rally on Corn got underway in May 2019 as Mars made its maximum declination. Mars at a declination low in early 2020 aligned to a price trend change. On the Wheat market, a Mars declination low in early 2020 aligned to a sharp rally which subsequently failed. In November 2019, an 80 cent per bushel rally aligned to Venus making its declination low. In the April-May period of 2020, a Venus declination high aligned to a steady decline in Wheat prices. The Venus declination high in the April-May period of 2020 aligned to a low and a trend change on Corn. In 2021, Venus was at its maximum declination from mid-May through mid-June. This represented the span of time that bridged a double top chart formation in Corn price. This span of time marked a decline in Wheat prices of over $1 per bushel. Mars made its declination maxima in mid-

April 2021. This marked the run-up in Wheat and Corn prices to levels not seen in 8 years. In early November, Venus recorded its minimum declination.

For 2022,

- ✿ Venus will be at its declination maximum around July 22

- ✿ Venus will be at its declination minimum around December 13

- ✿ Mars will be at its declination minimum around January 25 and at its maximum around December 5.

A deeper examination of declination reveals that in 1848 when the Chicago Board of Trade was founded, Mars was at its maximum declination and Moon was at 0 degrees declination. In 1877 when Wheat, Corn and Oats started trading, Venus was within about 3 degrees of its minimum declination.

Mercury Retrograde

Mercury retrograde plays a role in price pivot points on the grains. The price chart of Wheat futures in Figure 66 has been overlaid with Mercury retrograde events. Although not shown here, Mercury retrograde events do have a similar propensity to align to trend changes on Corn prices.

For 2022, Mercury will be:

- ✿ retrograde from January 14 through February 3

- ✿ retrograde from May 10 through June

- ✿ retrograde from September 9 through October 2.

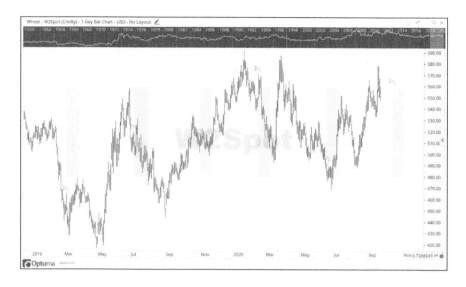

Figure 66
Wheat prices and Mercury retrograde events

Soybeans

Soybean futures started trading in Chicago on October 5, 1936. The horoscope in Figure 67 illustrates the planetary placements at that time. What is intriguing is the location of the Sun. Notice how it is exactly 90 degrees to the location of the Sun in the first trade horoscope for Wheat, Corn and Oats? Notice Sun is 180 degrees from the Sun in the 1848 CBOT natal chart? As I have previously suggested, the regulatory officials who determined these first trade dates knew more about astrology than we may think. Moreover, if one makes the reasonable assumption of a 7:00 am first trade, the Mid Heaven (MH) is at 14 Cancer.

Studying the 1877 natal horoscope for Corn and Wheat reveals a similarity to the 1936 Soybeans horoscope. Jupiter is at 18-19 degrees Sagittarius in both horoscopes.

Natal Transits

Events of transiting Sun making 0, 90, and 180-degree aspects to the natal Sun position in the 1936 first trade horoscope can be used to navigate the volatility of the Soybean market.

Figure 67
Soybeans first trade horoscope

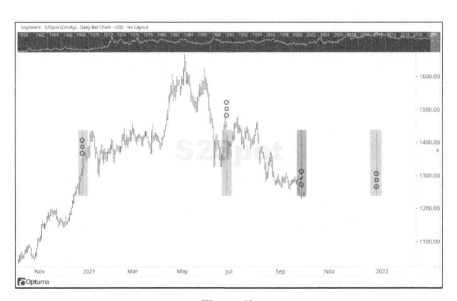

Figure 68
Soybeans and Sun/natal Sun events

Figure 68 illustrates the effect of transiting Sun making 0 and 90 degree aspects to the natal Sun position.

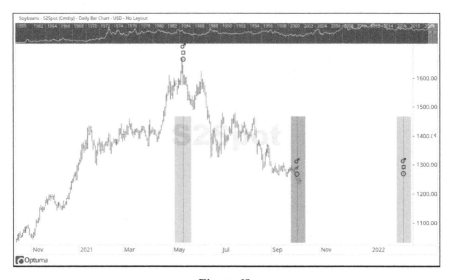

Figure 69
Soybeans and Mars/natal Sun events

Figure 69 illustrates the effect of Mars making aspects to the natal Sun location.

For 2022,

- ☼ transiting Sun will make a 90-degree aspect to natal Sun from January 1 through January 8

- ☼ A 90-degree aspect to natal Sun will occur from June 28 through July 11

- ☼ A 0-degree aspect will occur from September 26 through October 12.

For 2022,

- ☼ Mars will be 90 degrees to the 1936 natal Sun February 3 through February 16

○ Mars will be retrograde from October onwards and will not make any further aspects to natal Sun.

Mercury Retrograde

Mercury retrograde events also contribute to the price behavior of Soybeans. The Soybeans chart in Figure 70 illustrates the Mercury retrograde effect. If there is a trend change associated with Mercury retrograde, the trend change may come immediately beforehand, during or immediately afterwards. The use of a suitable chart technical indicator is essential to help identify the trend shifts.

For 2022, Mercury will be:

○ retrograde from January 14 through February 3

○ retrograde from May 10 through June 2

○ retrograde from September 9 through October 2.

Figure 70
Soybeans and Mercury events

Declination

Soybeans also have a tendency to record price trend changes in proximity to Venus recording maximum, and minimum declinations. The Soybean price chart in Figure 71 illustrates further. A price high and trend turning point in May 2021 aligns with a Venus declination maximum. A declination low in early-2021 caused a notable change in slope of the trend. Mars at its maximum declination also aligns to trend changes on Soybeans. Recall that the founding of the Chicago Board of Trade in 1848 had Mars at its maximum declination and Venus near its declination low.

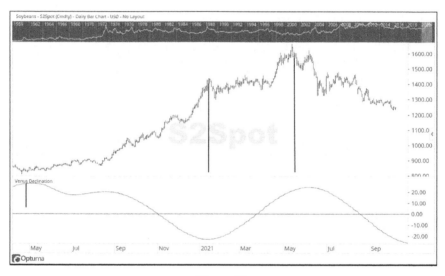

Figure 71
Soybeans and Venus Declination

For 2022,

☿ Venus will exhibit its maximum declination around July 22 and its declination low in mid-December

☿ Mars will be at minimum declination in mid-January and at maximum declination in early December.

136

Crude Oil

West Texas Intermediate Crude Oil futures started trading on a recognized exchange for the first time on March 30, 1983. A unique alignment of celestial points can be seen in the horoscope in Figure 72. Notice how Mars, North Node, (Saturn/Pluto/Moon) and Neptune conspire to form a rectangle.

Figure 72
Crude Oil First Trade horoscope

Natal Transits

Crude Oil can be a complex instrument to analyze using astrology. Given the peculiar rectangular shape that appears in the horoscope, one strategy for analyzing Crude Oil is to use natal transits with a focus on transiting Sun and transiting Mars making 0-degree aspects to the four corner points of the rectangle.

The tenor of the price reaction as Sun and Mars each pass the various corner points varies from year to year. Nevertheless, this strategy remains viable.

For 2022, the four corners of the peculiar rectangle will be passed by as follows:

- ✿ Sun will transit 0 degrees to the natal Mars location from April 9 through April 22

- ✿ Sun will transit 0 degrees to the natal Node location from June 7 through June 22

- ✿ Sun will transit 0 degrees to the natal (Saturn/Pluto/Moon) location from October 18 through November 1

- ✿ Sun will transit 0 degrees to the natal Neptune location from December 14 through the end of the year

- ✿ Mars will transit past the natal Mars location between June 22 and July 5

- ✿ Mars will approach the natal Node location in mid-October. Further progress will be stopped as Mars turns retrograde.

Transits of heliocentric Jupiter to the natal heliocentric Jupiter location in the 1983 first trade horoscope are also deserving of attention. In 2015, Oil prices fell over $20 per barrel as heliocentric Jupiter passed 90 degrees to the natal heliocentric Jupiter point. Oil prices fell over $30 per barrel in late 2018 as heliocentric Jupiter passed 0 degrees to the natal heliocentric Jupiter point. The strong rally in Oil prices in late 2021 started in earnest in late August as heliocentric Jupiter approached 28 Aquarius and a 90-degree aspect.

Retrograde

Crude Oil is influenced by Mercury retrograde and Venus retrograde.

For 2022, Mercury will be:

- ✿ retrograde from January 14 through February 3

- ✿ retrograde from May 10 through June 2

- ✿ retrograde from September 9 through October

☼ For 2022, Venus will be retrograde during January.

Declination

At the 1983 first trade date, Moon was within 5 degrees of being at its 0- degree declination point. A look at a Crude Oil chart suggests that monthly times of Moon at or near 0 declination are potential times for trend changes.

Cotton

Figure 73
Cotton futures First Trade horoscope

After sifting through back-editions of New York newspapers, I have come to conclude that Cotton futures first started trading on June 20, 1870. The horoscope wheel in Figure 73 illustrates planetary placements at that time. At first glance, I find it peculiar that the Moon is at the same degree and sign location (24 Pisces) as is the Mid-Heaven in the New York Stock Exchange natal horoscope wheel from 1792. If one assumes that at 6:00 am a trade was initiated, the Mid-Heaven (MH) point is at 24 Pisces. Furthermore, Mars, Jupiter and Mercury are clustered around the location 9 Gemini which is where one finds

Uranus in the USA 1776 natal chart. Surely the selection of this Cotton first trade date is no accident.

Natal Transits

Events of transiting Sun passing 0 and 90 degrees to the natal Sun position are an effective tool for traders to use when navigating the choppy waters of Cotton prices. The Cotton price chart in Figure 74 illustrates further. Note the alignment to lows in March 2021 and the alignment to the start of a sharp rally in September.

Figure 74
Cotton futures and Sun/natal Sun aspects

In 2022, transiting Sun will aspect natal Sun as follows:

- ☼ Transiting Sun will pass 90 degrees to natal Sun from March 7 through March 23

- ☼ Transiting Sun will pass 0 degrees conjunct to natal Sun from June 10 through June 30

- ☼ Transiting Sun will pass 90 degrees to natal Sun from September 13 through September 30

✿ Transiting Sun will pass 180 degrees to natal Sun from December 12 through December 28.

Figure 75
Cotton futures and Venus/natal Moon aspects

Venus and Natal Moon

One other astro phenomenon traders may wish to consider as a tool to use is the occurrence of Venus passing by the natal Moon position at 24 Pisces.

Not a frequent event, it is nonetheless one to pay attention to. The price chart in Figure 75 illustrates.

During 2020, Venus transited past the natal Moon from late January through early February. This passage triggered a 20 cent per pound decline in cotton prices. A 3 cent move on Cotton futures is $1500. For 2021, Venus passed the 24 Pisces point from March 12 through March 21. Days later, a rally unfolded that took Cotton from the 77 cents per pound level to the 91-cent level.

✿ In 2022, Venus will transit past the 1870 natal Moon late April through early May.

Declination

At the 1870 first trade date, Mars was very near its declination maximum. A back-study of Cotton prices shows a good correlation between trend changes and Mars declination maxima.

☼ For 2022, Mars will be at its maximum declination in early December.

Coffee

Figure 76
Coffee futures First Trade horoscope

Coffee futures started trading in New York in early March of 1882. The horoscope wheel in Figure 76 illustrates planetary placements at that time.

Natal Transits

In the Coffee horoscope, note the 180 degree aspect between Sun and Uranus. Louise McWhirter in her 1937 writings cautioned it is not wise to invest in situations where this sort of aspect exists because one will experience many wild ups and downs in price over time. A quick look at a 10-year price chart of Coffee reveals a price range of $0.65 per pound to $3.06 per pound with many wild swings. Your point is well taken, Ms. McWhirter.

Coffee prices react to transits of Sun passing 0 and 90 degrees to the natal Sun position at 16 Pisces. More importantly, however, at the 1882 first trade date Venus, Moon and Sun were all at or near 0 declination. Mars was at its maximum declination. A look back in time across coffee prices shows a clear alignment of trend changes to Venus being near 0 declination. Likewise for Mars being at or near its maximum declination.

In 2022,

- ✿ Transiting Sun will pass 0 degrees conjunct to natal Sun from February 29 through March 16

- ✿ Transiting Sun will pass 90 degrees conjunct to natal Sun from May 31 through June 17

- ✿ Transiting Sun will pass 180 degrees to natal Sun from August 31 through September 19

- ✿ Transiting Sun will pass 90 degrees to natal Sun from December 10 through December 16

- ✿ Venus will be near 0 degrees declination around May 5 and again at October 1, 2022.

- ✿ Mars will be at maximum declination in early December.

Non-natal Transits

The positioning of Sun opposite Uranus in the 1882 natal horoscope is intriguing. 180-degree aspects between transiting Sun and Uranus can be used to further assist the Coffee trader with decision making.

In October 2019, a 180-degree aspect aligned to a trend change as Figure 69 shows. A similar such occurrence came again in October 2020. At this time of writing, Coffee prices have been pushing higher, but late October will deliver another Sun/Uranus opposition.

Figure 77
Coffee prices and Sun/Uranus aspects

For 2022, Sun will make the following hard aspects (0, 90, 180 degree) with Uranus:

- ☼ 90 degrees in late January

- ☼ 0 degrees in late April

- ☼ 90 degrees in late July

- ☼ 180 degrees in late October.

Sugar

Figure 78
Sugar Futures first trade horoscope

Sugar as a bulk commodity started trading in New York as early as 1881. Old editions of New York newspapers suggest that in September 1914 there were plans to open a formal Sugar Exchange, but these plans were scuttled by World War I. Following the war, a formal Exchange did open on August 31, 1916. The horoscope wheel in Figure 78 illustrates planetary placements at that time. What stands out on this chart wheel is the T-Square formation with Mars at its apex. As well, the Mid Heaven (MH) is at 14 of Cancer is one assumes that the very first trade was conducted at 8:20 am.

Natal Transits

Events of transiting heliocentric Mars making 0 and 90-degree aspects to the natal heliocentric Mars location at 22 Scorpio have a good propensity to align to pivot swing points. This is the intriguing part of astrology. Even though a significant feature such as a T-square might be noted on a geocentric horoscope, sometimes it is a heliocentric event that triggers action.

For 2022, transiting heliocentric Mars will make the following aspects to heliocentric natal Mars:

☼ Completing a 0-degree aspect in early January.

☼ 90 degrees in mid-July.

Mercury Retrograde

Mercury retrograde events have a propensity to align to short-term trend changes on Sugar price.

For 2022, Mercury will be:

☼ retrograde from January 14 through February 3

☼ retrograde from May 10 through June 2

☼ retrograde from September 9 through October 2.

Declination

At the 1916 first trade date, Mercury was within 2 degrees of being at its 0-degree declination point. A look back at past price data for Sugar shows that this connection still holds.

☼ For 2022, Mercury will be passing through 0 degrees of declination April 1, August 22, September 27, and October 15.

Cocoa

Figure 79
Cocoa futures first trade horoscope

Cocoa futures started trading in New York in early October 1925. The horoscope in Figure 79 shows planetary placements at the first trade date. What I find peculiar on this horoscope wheel is the Mid-Heaven point located at 14 of Cancer, that same mysterious point that appears in the First Trade horoscope of the New York Stock Exchange.

Mercury Retrograde

Mercury retrograde events have a high propensity to align to pivot swing points on Cocoa price. The price chart in Figure 80 illustrates further. Sometimes a Mercury retrograde event can deliver some erratic volatility as part of an ongoing trend, while other times the retrograde event can bring about a complete change of trend either immediately before the retrograde or immediately after. Either way, Mercury retrograde events deserve close scrutiny.

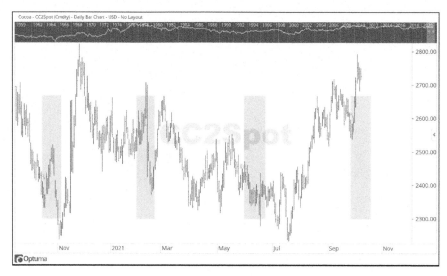

Figure 80
Mercury retrograde and Cocoa price

The alignment to price swing points is very evident in studying Figure 112.

For 2022, Mercury will be:

- ☿ retrograde from January 14 through February 3

- ☿ retrograde from May 10 through June 2

- ☿ retrograde from September 9 through October 2.

Conjunctions and Elongations

The 1925 natal horoscope shows Sun and Mercury conjunct (0 degrees apart). My research has shown that events of Mercury being at its maximum easterly and westerly elongations and events of Mercury being at its Inferior and Superior conjunctions align quite well to pivot swing points. The price chart in Figure 81 illustrates both phenomena further. The East and West elongation alignment to pivot price points is intriguing.

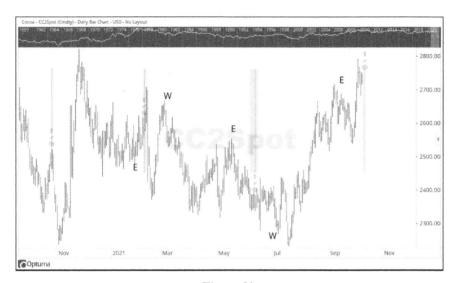

Figure 81
Cocoa price and Mercury cycles

For 2022:

- ✪ Mercury will be at its greatest easterly elongation January 7, April 29, August 27, and December 21

- ✪ Mercury greatest westerly elongation events will occur February 17, June 16, and October 9.

Declination

At the 1925 first trade date, Sun, Moon and Mars were all at or near their 0-degree declination levels. A look back at past price data for Cocoa shows that this connection still holds.

For 2022:

- ✪ Mars will be 0-degrees declination level around June 1.

Feeder Cattle

Figure 82
Feeder Cattle first trade horoscope

Feeder Cattle futures started trading in Chicago on November 30, 1971. The horoscope in Figure 82 shows planetary placements at the first trade date. What is interesting is Mercury and Venus being conjunct within 4 degrees of each other. Traders of Feeder Cattle futures may wish to anticipate short term trend changes by identifying times when Mercury and Venus come into conjunction. Another interesting feature is the 90- degree square aspect between Mars and Jupiter. A look back at past chart data suggests potential for short term trend swings when these planets are at 90-degree and 0-degree aspects. Lastly, (assuming a 9:00 start to trading) I doubt it was an accident to have the Immum Coeli point (bottom of horoscope wheel) aligned to 24 Pisces, the natal Mid-Heaven point of the NYSE from 1792. A look back at past charts to the 1980s shows that times of Venus passing 24 Pisces often align to short term trend swings on Feeder prices.

Declination

At the 1971 first trade date, Venus and Mercury were very close to their minimum declinations. A look back at past chart patterns shows that Venus maximum and minimum declinations bear a good alignment to points of short-term trend change on Feeder prices. There is decent correlation to Mercury declination minima as well, but Venus is the more robust candidate.

For 2022:

- ☿ Venus will be at maximum declination around July 22

- ☿ Venus will be at minimum declination around December 11.

Live Cattle

Figure 83
Live Cattle first trade horoscope

Live Cattle futures started trading in Chicago on November 30, 1964. The horoscope in Figure 83 shows planetary placements at the first trade date. The most puzzling feature of this horoscope is that it comes at the same day (November 30) as the Feeder Cattle first trade date. Again, note the alignment of the IC point to 24 Pisces. A look back at past charts to the 1960s shows that times of Venus passing 24 Pisces often align to short term trend swings on Live Cattle prices.

Declination

At the 1964 first trade date, Mercury was at its declination minimum, Venus was its minimum declination, and Moon was near 0-degrees declination. A look back at past chart patterns shows that Venus maximum and minimum declinations bear a good alignment to points of short-term trend change on Live Cattle prices. There is decent correlation to Mercury declination minima and 0-degree points as well, but Venus is the more robust candidate.

For 2022:

☼ Venus will be at maximum declination around July 22

☼ Venus will be at minimum declination around December 11.

Lean Hogs

Figure 84
Lean Hogs first trade horoscope

Lean Hog futures started trading in Chicago on February 28, 1966. The horoscope in Figure 84 shows planetary placements at the first trade date. The most illuminating feature of this horoscope is the appearance of a large square pattern (assuming a 6:45 am trade start) with corners at Saturn, Jupiter, Pluto, and the Mid-Heaven. A look back at past chart data shows a decent alignment to price swing points when Mars and Venus pass the corner points of the square. Be cautioned, however, that Lean Hogs are extremely volatile and not well-suited for risk-averse traders.

Declination

At the 1966 first trade date, Mercury was within 1 degree of being at 0 degrees declination. Mars was within 3 degrees of being at 0 degrees declination. A look back at past chart patterns shows a decent alignment between Mercury 0-degree declination points and price trend swings. A similar observation holds for Mars. But these alignments do not erase the fact that Hogs are volatile to trade.

For 2022:

- ✧ Mars will be at minimum declination in late January
- ✧ Mars will be at maximum declination in early December
- ✧ Mars will pass 0 degrees declination around June 1
- ✧ Mercury will exhibit 2 maxima points at May 5 and July 11
- ✧ Mercury will exhibit minimum declination in early December
- ✧ Mercury will pass through 0 degrees declination on March 30, August 22, September 27 and October 14.

Lumber

Figure 85
Lumber first trade horoscope

Lumber futures started trading in Chicago on October 1, 1969. The horoscope in Figure 85 shows planetary placements at the first trade date. Assuming a 7:15 am first trade transaction, the Mid Heaven point is at 14 Cancer.

Mercury Retrograde

The most illuminating feature of this chart is the fact that Mercury was retrograde. If trading lumber futures holds appeal for you, pay close attention to Mercury retrograde events as they might well align to price swing points. In May 2021, Lumber made headline news as prices reached level never before seen at near $1400. Within two weeks of this peak, Mercury turned retrograde and with that a swift price drawdown began.

For 2022, Mercury will be:

☿ retrograde from January 14 through February 3

◇ retrograde from May 10 through June 2

◇ retrograde from September 9 through October 2.

Declination

At the 1969 first trade date, Mars was at its declination minimum. Moon was at its declination maximum. Mercury was within a couple degrees of being at 0-degrees declination. Looking back at the historically significant price high in May 2021 shows that Mars was at its declination maximum (the opposite extreme from the 1969 first trade date), Moon was at 0 declination, and Mercury was at its declination maximum.

For 2022:

◇ Mars will be at minimum declination in mid-January and at maximum declination in early December

◇ Mercury will pass through 0 degrees declination April 1, August 22, September 27 and October 15

◇ Mercury will exhibit 2 maxima points at May 5 and July 11

◇ Mercury will exhibit minimum declination in early December.

Platinum

Figure 86
Platinum first trade horoscope

Platinum futures started trading in New York on December 3, 1956. Assuming an 8:50 am first trade transaction, note the alignment of the IC point to 24 Pisces.

Declination

At that date, Moon was at its minimum declination and Mercury was not only at a declination low, it was what astrologers call out-of-bounds (OB). A planet is OB when its declination maximum or minimum exceeds by a few degrees what its maxima or minima normally would be.

Looking back at past chart patterns shows a significant price run-up in Platinum that took prices from $1070 in 2007 to $2230 in 2008. This sizeable rally started at a Mercury minima event and the double top formation ended at a Mercury maxima event. Price subsequently declined to $800 by late 2008. The low point of this sell-off was reached at a Mercury minimum event.

For 2022:

- ☼ Mercury will pass through 0 degrees declination April 1, August 22, September 27 and October 15

- ☼ Mercury will exhibit 2 maxima points at May 5 and July 11

- ☼ Mercury will exhibit minimum declination in early December.

Palladium

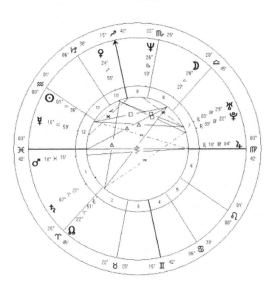

Figure 87
Palladium first trade horoscope

Palladium futures started trading in New York on January 22, 1968.

Declination

At that date, Moon was within a degree of being at 0-degrees declination. Venus was at its minimum declination. A look back at past chart patterns shows that Venus declination minima do figure in the various rallies that have unfolded over the years.

For 2022:

☼ Venus will be at maximum declination around July 22

☼ Venus will be at minimum declination around December 11.

Natural Gas

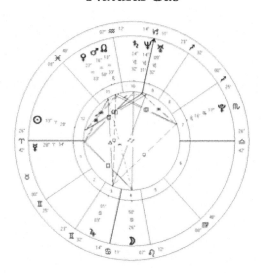

Figure 88
Natural Gas first trade horoscope

Natural Gas futures started trading in Chicago on April 3, 1990. At that date, Moon had just made its declination maximum. No other planetary declination level stands out as being curiously interesting. The above horoscope does however have that always intriguing 14 of Cancer point at the horoscope IC point. A look back at past chart data suggests that times when Mercury, Mars, Venus, and Sun variously transit past 14 Aries, 14 Cancer, 14 Libra, and 14 Capricorn often align to price swing points.

CHAPTER THIRTEEN

Gann, Astrology & Weather

Tunnel Through the Air

In W.D. Gann's 1927 book *Tunnel Through the Air*,[1] central character Robert Gordon places trades on Wheat, Corn and Cotton in accordance with pending weather patterns he seems to know about. If the reader generates a horoscope for these various trade dates, aspects between some of the larger outer planets are plainly visible in the zodiac wheel. From this, we can infer that Gann was aware that aspects between these outer planets could affect the weather and therefore commodity prices. What follows in this brief chapter is a summary overview of Gann's ideas on weather. Although 2022 is not 1927, and weather patterns today are more volatile that they were 90 years ago, understanding Gann's connection between planetary aspects and weather is valuable information to have nonetheless.

Uranus and Neptune

Taking cues from Gann's book, it appears that times when Uranus and

Neptune make aspects to one another, particularly in January through April, foretell a wet weather period ahead. Wet weather can delay crop planting dates which can impact crop yields and thereby price.

Using *Solar Fire Gold* software, a search for times when these two planets were at any geocentric aspect, reveals a 45-degree aspect during the first 8 months of 2018. For most of this timeframe, Neptune was in the sign of Pisces, a water sign. A 45-degree aspect is also generally evident up to May 2020 with Neptune also in the sign of Pisces. As Neptune turns direct in late 2020, the 45-degree aspect will briefly re-appear until about March 2021. After March, the aspect widens and starts to approach 50 degrees.

The National Agricultural Statistics Service notes that in 2018: *hurricanes Florence and Michael brought destructive damage to coastal areas and also left a footprint inland which affected crops. In many areas, ample rainfall accompanied the above-normal temperatures, leading to record-high yield expectations for the Nation's corn and soybeans. As the growing season ended, rainfall intensified in many areas, contributing to substantial harvest delays. Fieldwork further slowed in some areas in November, when cold, stormy weather hampered final harvest efforts and slowed winter wheat planting, emergence, and establishment across portions of the Plains, Midwest, South, and East.* [2]

The National Agricultural Statistics Service notes that in 2019:

Despite the mid-year demise of El Niño, wet conditions were a hallmark of the 2019 growing season across large sections of the Plains and Midwest. Midwestern wetness, which intensified in March and persisted through spring and into summer, caused significant planting delays and subsequently slowed the development of crops such as corn and soybeans. Planting and developmental delays extended to other crops, including sugarbeets, sunflowers, and spring wheat. [3]

It seems fair to conclude that Uranus/Neptune aspects do have an effect on weather patterns.

If using these aspects as a trading tool, one must watch the price trend for evidence of a technical pattern breakout. In this era of climate change, it can no longer be assumed that these aspects will automatically deliver excess moisture. Moreover, even if moisture does result, it may not be adverse over an entire growing region. As well, crops today have been genetically modified so that even with a planting delay or even with a bit too much moisture, a decent harvest can still result. Farmers today also tend to hold back some product in the bin after a harvest. So what might have given Gann some notable price surges in 1927, may not deliver the same today.

Jupiter and Neptune

According to Gann's book, it appears that times when Jupiter and Neptune make aspects to one another, particularly in April through June, foretell a desirable weather period ahead. Desirable growing weather can result in above forecast yields and lower prices come harvest time.

Using *Solar Fire Gold* software, a search for times when these two planets were at any geocentric aspect revealed a 120-degree aspect in the February-April period of 2014, a 180-degree aspect in the March-July period of 2016, a 120-degree aspect in the April-September period of 2018, a 90-degree aspect in the 2019 growing season, and a 60-degree aspect in the 2020 growing season.

Crop statistics show that weather patterns were not uniform in 2014, but on balance soil moisture conditions did improve over the year. Data for 2016 shows yields on the year were up from 2015. The 120-degree aspect in 2018 conflicts with the Uranus/Neptune aspect of 2018. A positive planetary influence overlain on a negative influence might explain why price increases on futures contracts were not bigger. The 90-degree aspect in 2019 conflicts with the Uranus/Neptune aspect of 2019. It appears that the Uranus/Neptune duo won the weather battle.

The Spring of 2021 saw the aspect between these planets diminish to less than 30 degrees. This opened the door for another aspect pattern, this one involving Saturn.

Saturn and Neptune

Taking another cue from Gann's book, it appears that times when Saturn and Neptune make aspects to one another, can promote dry weather which will have a positive impact on prices come harvest time.

For 2021, the planting season in April and May was characterized by Saturn and Neptune drawing into a 45-degree aspect. Neptune was in the 'wet' sign of Pisces, but Saturn was in the 'dry' sign of Aquarius. It appears that Saturn prevailed in the extreme. Crops fared poorly across western Canada and the US mid-west. Many of the older farmers I talked in Saskatchewan in 2021 claimed that based on stories they had heard while growing up, 2021 was on par with 1936 as one of the worst crop years.

As I finish this manuscript, Saturn continues to prevail. Weather conditions remain unseasonably dry. Looking ahead to the seeding time of April-May, 2022 Saturn and Neptune will be at a 30-degree aspect. If Saturn continues to prevail, the dry conditions might persist. While this might bode well for grain prices, if soil moisture continues to lack, the 2022 crop will again be negatively affected. Farm operators who cannot generate a crop cannot benefit from higher commodity prices.

CHAPTER FOURTEEN
Price, Time, & Quantum Lines

Gann Fan Lines

Gann Fan Lines are a technique in which a starting point of a significant high or low is selected. From this point, angles (vectors) are projected outwards. These vectors are the 1x1, 1x2, 1x4, 1x8, 2x1, 4x1 and 8x1. In and of themselves, these Gann Lines are not related to astrology. However, in my opinion, they should be applied to charts and used in combination with astrology.

Many market data software platforms will come with a Gann Fan function already built in. The confusion with Gann lines comes from the mathematical method of constructing the lines. In fact, in the *Optuma/ Market Analyst* program there are no fewer than ten ways to apply Gann Lines to a chart. If Mr. Gann were around today he would probably shake his head in bewilderment at how convoluted his technique has become. My preference for applying Gann Lines is the methodology used by Daniel Ferrera in his book *Gann for the Active Trader.* [1] Ferrera's method is based on Gann Square of Nine mathematics.

To illustrate the creation of Gann Fan Lines, I will use Gold as an example. Follow the examples in this chapter and you will be able to superimpose Gann Fan lines on any chart you wish using a pencil and a ruler:

On August 16, 2018 Gold made a price low at $1175. This is the point from which I wish to extend Gann lines.

Step 1: Take the $1175 figure and express it as the number 1175. Take the square root of 1175 and you get 34.27. This will be your time factor.

Step 2: Add 1 to 34.27 and re-square this figure to get 1244.

Step 3: We can now state that our time factor is 34.27 calendar days. For simplicity, we can round this off to 34 days. We can further state that our price factor is 1244 minus 1175 = $69.

Step 4: From the August 16 date, extend a line so that it passes through the time co-ordinate (August 16+34 days = September 20) and the price co-ordinate $1244 ($1175 + $69 = $1244). This line is the Gann 1x1 line.

Step 5: From the August 16 date, extend another line so that it passes through the time co-ordinate (August 16 + (34 x 2) days = October 24) and the price co-ordinate $1244. This line is the Gann 1x2 line.

Step 6: From the August 16 date, extend another line so that it passes through time co-ordinate ((August 16+ (4 x 34) days = January 3) and the price co-ordinate $1244. This line is the Gann 1x4 line.

The Gold price chart in Figure 89 has these Gann lines overlaid starting from the August $1175 low.

Figure 89
Gann Lines applied to a Gold Chart

Notice from the $1175 low point, the price high in July 2020 was capped by the 1x2 line. The 1x4 line is currently overhead resistance at this time of writing.

As one more example, consider Wheat which registered a significant low on April 30, 2019 at $4.16 per bushel.

Step 1: Take the $4.16 value and express it as the number 4160. Take the square root of 4160 and you get 64.49. This will be the time factor.

Step 2: Add 1 to 64.49 and re-square this figure to get 4290. Shifting the decimal over we get $4.29.

Step 3: We can now state that the time factor is 64 calendar days. We can further state that our price factor is $4.29 minus $4.16 = $0.13.

Step 4: From the April 30 date, extend a line so that it passes through the time co-ordinate (April 30 + 64 days = July 4) and the price co-ordinate $4.29. This line is the Gann 1x1 line.

Step 5: From the April 30 date, extend another line so that it passes through the time co-ordinate (April 30 + 2 x 64 days = September 8) and the price co-ordinate $4.29. This line is the Gann 1x2 line.

Step 6: From the April 30 date, extend another line so that it passes through the time co-ordinate (April 30 + (0.33) x 64 days = May 21) and the price co-ordinate $4.29. This line is the Gann 3x1 line.

The Wheat price chart in Figure 90 has these Gann lines overlaid starting from the April 30 low. The price high in May 2021 aligns to the 8x1 line. Twice in 2021 the 4x1 line provided support.

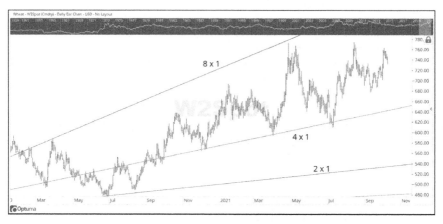

Figure 90
Gann Lines applied to a Wheat Chart

Price Square Time

The concept of *price square time* says that at a significant swing point and trend change on a stock chart, commodity price chart or index chart exists because price and time have squared with one another. That is, a planet has advanced a certain number of degrees and price has changed by that same number of degrees or a multiple of those degrees.

To illustrate, Figure 91 presents a segment from a Gold futures chart dating to 2018-19.

Price Rise = $187

Figure 91
Gold price and the concept of Price Square Time

This chart shows that from August 18 through February 2019 there was a clear uptrend, which broke in February 2019. The price rise from low to high (intraday) was $187.

During this timeframe, Sun advanced 187 degrees (23 Leo to 0 Pisces) and Venus advanced from 8 Libra to 12 Capricorn (94 degrees; 94 x 2 = 188), Mars moved from 29 Capricorn to 2 Taurus (93 degrees; 93 x 2 = 186). Mercury and Jupiter moved 210 degrees and 35 degrees respectively, so they are not significant to this argument.

The planetary movement of Sun, Venus and Mars squares (closely matches) the price movement of Gold. Identifying price square time events will take some work on your part, but, the results will make the effort worthwhile. When dealing with large price values, such as the Dow Jones, the Nasdaq or the S&P 500, you will use a multiplier of the planetary movement values. To illustrate, consider the panic sell-off surrounding the COVID situation in early 2020.

From the February 12, 2020 trend turning point to the March 23 panic low point, the S&P 500 dropped 1149 points (close to close basis).

During this same time frame, heliocentric Venus moved 23 degrees. Take 50 times 23 and the result is 1150. Thus, one can argue that price and time had squared at the March 2020 lows.

During this same time frame, heliocentric Mars moved 16 degrees. Taking 72 times 16 gives a result of 1152. Thus, one can argue further argue that price and time had squared at the March lows.

I have yet to fully comprehend the mysteries of price squaring with time. Why in one instance does a 20 X multiplier work, when in the other instance a 72 X multiplier is needed? Why are these multipliers always whole integer numbers?

If you are prepared to take the time to determine both geocentric and heliocentric movements of Sun, Venus and Mars, Price Square Time will prove to be a powerful tool for you.

As two final examples, consider the Dow Jones Average and the Nasdaq 100.

From the March 2020 lows to what appears to be a trend change at August 21 2021, the Dow Jones Average advanced 13,484 points on a close to close basis or 17418 points on a low to high basis. Meantime, heliocentric Mars advanced 273 degrees and Venus a total of 823 degrees. The 13,484 point advance shows that Mars advanced a multiple of 49.4 (50 X in round figures) times. Price and time had squared. The 17,418 point move shows that Venus advanced a multiple of 21 times. Again, price and time had squared.

From an orderly pullback in March 2021, the Nasdaq advanced to reach a peak on September 7. This advance (close to close) is 10 X the advance of heliocentric Venus and a 35 X advance of heliocentric Mars. Both are whole integers. Price and time have squared.

Newton and Einstein

In the early 1700s, scientist Sir Isaac Newton developed his *Theory of Universal Gravitation* in which he said planets in our solar system are attracted to one another by gravity. Newton further said that space and time were absolute and that the world functioned according to an absolute order. Furthermore, he said that space was a three-dimensional entity and time was a two-dimensional entity.

In the early 1900's, Albert Einstein advanced his *Theory of Relativity* that posited Newton's absolute model was outdated. Einstein said the passage of time of an object through space was related to its speed with respect to that of another observer. Thus was penned the concept of relative space-time in which space was not uniform.

Einstein further stated that relative space-time could be distorted depending on the density of matter. That is, space-time in the area of the Sun is more distorted because the Sun has a great, huge mass. Light particles travelling near the Sun are then distorted from their linear path due to the mass of the Sun.

Quantum Price Lines

Quantum Price Lines are based on Einstein's quantum theory. The notion of Quantum Lines posits that the price of a stock, index or commodity can be thought of as a light particle or electron that can occupy different energy levels or orbital shells.

Author and market researcher Fabio Oreste combined the notion of quantum price lines with Einstein's theory that the fabric of space-time can be bent. Picture a group of people holding the edges of a large blanket. They pull on the edges until the blanket is stretched tight. Next, someone places a ball on the tight blanket. The weight of the ball causes a slight sag in the blanket fabric. Oreste says the point of maximum curvature of the sagged portion is akin to a Quantum Price Line. In his book entitled *Quantum Trading*,[2] Oreste details his formula for Quantum Price Line calculation:

Quantum Line = (N x 360) + PSO ;

Where PSO = heliocentric planetary longitude x Conversion Scale

Where N is the harmonic level = 1,2,3,4,5,6,8,…

Always think of harmonics in terms of divisions of a circle. A 3rd harmonic (N=3) is 120 degrees. A 5th harmonic is 360/5 =72 degrees. And so on.

Where Conversion Scale = 2n ; 1,2,4,8,16,…; where n=0,1,2,3,4,….

When dealing with prices less than 360, the inverse variation of the formula is used.

Quantum Line = (1/N x 360) + PSO

The technique then allows one to calculate various sub-divisions of these Quantum Lines. Taking the value of the calculated Quantum Line, one would generate the sub-divisions by multiplying by 1.0625, 1.125, 1.875, 1.25, etc… in steps of 0.0625.

Please note the use of *heliocentric* planetary data in these Quantum Line calculations. There are websites that will provide you with this data such as **www.astro.com/swisseph.**

Alternatively, you can find a Heliocentric Ephemeris book such as *The American Heliocentric Ephemeris, 2001-2050.*

To assist you with calculating Quantum Lines, consider the following example:

On a given date, suppose the following heliocentric planetary positions are noted: Mars 306 degrees, Jupiter 307 degrees, Neptune 324 degrees, Pluto 271 degrees.

In this example, let N=1 and let the conversion scale be set to CS=1.

PSO will be the planetary longitude x CS.

The Oreste point of maximum curvature for these planets is then:

Mars: (N x 360) + PSO; which is (1 x360) + 306 = 666
Jupiter: 360 + 307 = 667
Neptune: 360 + 324 = 684
Pluto: 360 + 271 = 631

If you were to take another date in the future and calculate the points of maximum curvature, you could then join the two points for each planet. By definition two points joined equals a line. You could extend these lines out into the future. These lines are called Quantum Lines (or QL's).

If the above maximum curvature numbers seem oddly familiar, that is because they are. The S&P 500 March 6th, 2009 lows delivered an intra-day low of 665.7 and on the day the close was 687. Indeed. Mars, Jupiter and Neptune all acted in concert in March 6, 2009 to provide a floor of support under the US equity market.

For larger price points such as those in an equity index, it becomes necessary to use a bigger value for N. Consider the following example of the S&P 500. The high on the S&P 500 at September 7, 2021 was 4546. Intuitively, a larger number such as this will demand something more than N=1 and CS=1. The question is, are there Oreste points of maximum curvature at or near this level that would suggest the S&P 500 hit a significant resistance level?

The first step in answering this question is to obtain the heliocentric planetary data for the date in question. Positions of planets were:

Mars at 180 degrees, Jupiter at 328, Neptune at 351, Saturn at 311, Pluto at 295.

Next, choose a CS value. Given the magnitude of the S&P 500, I am going to pick CS=2^3 = 8.

The formula is: (N x 360) + (heliocentric position x conversion scale)

Calculating the back half of the formula gives us:

Mars: 8*180 = 1440
Jupiter: 8*328 = 2624
Neptune: 8*351 = 2808
Saturn: 8*311 = 2488
Pluto: 8*295 = 2360

Next, a harmonic N value is required. Consider the 5th harmonic of N=5.

The formula then yields:

Mars: (5x360) + 1440 = 3240
Jupiter: (5x360) + 2624 = 4424
Neptune: (5x360) + 2808 = 4608
Saturn: (5x360) +2488 = 4288
Pluto: (5x360) + 2360 = 4160

From these numbers we can conclude that a Neptune 5th harmonic at 4608 was foretelling of overhead resistance on September 7, 2021.

Next, consider the 6th harmonic and N=6x360 = 2160.

The recalculated values become:

Mars: (6x360) + 1440 = 3600
Jupiter: (6x360) + 2624 = 4784
Neptune: (6x360) + 2808 = 4968
Saturn: (6x360) + 2488 = 4648
Pluto: (6x360) + 2360 = 4520

From these numbers we can conclude that Pluto 6th harmonic at 4520 was acting as resistance on September 7th. A Saturn 6th harmonic was situated close by.

What follows is a suggested list of some points of maximum curvature you can apply to various indices and commodities for 2022. These various points are based on the following heliocentric planetary positions at January 1, July 1, and December 30, 2022. As 2022 starts, sketch the Jan 1 and July 1 points on your charts. Join the points with a line. Later in 2022, add the December 30 point and extend your lines.

Heliocentric Degree Positions

HELIOCENTRIC DEGREE POSITION			
Planet	Jan 1, 2022	July 1, 2022	Dec 30, 2022
Jupiter	339	356	372
Saturn	314	320	325
Neptune	352	353	354
Uranus	43	45	47
Pluto	296	297	298

2022 Quantum Levels

S&P 500 Index (CS=8)

PLANET & HARMONIC	JAN 1	JULY 1	DEC 30
Jupiter 5th	4512	4648	4776
Jupiter 6th	4872	5008	5136
Jupiter 7th	5232	5368	5496
Saturn 5th	4312	4360	4400
Saturn 6th	4672	4720	4760
Saturn 7th	5032	5080	5120
Neptune 5th	4616	4624	4632
Neptune 6th	4976	4984	4992
Neptune 7th	5336	5344	5352
Pluto 5th	4168	4176	4184
Pluto 6th	4528	4536	4544
Pluto 7th	4888	4896	4904

Nasdaq Composite Index (CS=32)

PLANET & HARMONIC	JAN 1	JULY 1	DEC 30
Jupiter 10th	14448	14992	15504
Jupiter 12th	15168	15712	16224
Jupiter 16th	16608	17152	17664
Saturn 10th	13648	13840	14000
Saturn 12th	14368	14560	14720
Saturn 16th	15808	16000	16160
Neptune 10th	14864	14896	14928
Neptune 12th	15584	15616	15648
Neptune 16th	17024	17056	17088
Pluto 10th	13072	13104	13136
Pluto 12th	13792	13824	13856
Pluto 16th	15232	15264	15296

FTSE 100 Index (CS=16)

PLANET & HARMONIC	JAN 1	JULY 1	DEC 30
Jupiter 4th	6864	7136	7392
Jupiter 5th	7224	7496	7752
Jupiter 6th	7584	7856	8112
Saturn 4th	6464	6560	6640
Saturn 5th	6824	6920	7000
Saturn 6th	7184	7280	7360
Neptune 4th	7072	7088	7104
Neptune 5th	7432	7448	7464
Neptune 6th	7792	7808	7824
Pluto 4th	6176	6192	6208
Pluto 5th	6536	6552	6568
Pluto 6th	6896	6912	6928

S&P ASX 200 Index (CS=16)

PLANET & HARMONIC	JAN 1	JULY 1	DEC 30
Jupiter 4th	6864	7136	7392
Jupiter 6th	7584	7856	8112
Jupiter 8th	8304	8576	8832
Saturn 4th	6464	6560	6640
Saturn 6th	7184	7280	7360
Saturn 8th	7904	8000	8080
Neptune 4th	7072	7088	7104
Neptune 6th	7792	7808	7824
Neptune 8th	8512	8528	8544
Pluto 4th	6176	6192	6208
Pluto 6th	6896	6912	6928
Pluto 8th	7616	7632	7648

Gold Futures (CS=1)

PLANET & HARMONIC	JAN 1	JULY 1	DEC 30
Jupiter 4th	$1779	$1796	$1812
Jupiter 5th	$2139	$2516	$2532
Jupiter 6th	$2499	$3236	$3252
Saturn 4th	$1754	$1760	$1765
Saturn 5th	$2474	$2480	$2485
Saturn 6th	$3194	$3200	$3205
Neptune 4th	$1792	$1793	$1794
Neptune 5th	$2512	$2513	$2514
Neptune 6th	$3232	$3233	$3234
Pluto 4th	$1736	$1737	$1738
Pluto 5th	$2456	$2457	$2458
Pluto 6th	$3176	$3177	$3178

Silver Futures (CS=1/64, N=16,18,20)

PLANET & HARMONIC	JAN 1	JULY 1	DEC 30
Jupiter 16th	$27.80	$28.06	$28.31
Jupiter 18th	$25.30	$25.56	$25.81
Jupiter 20th	$23.30	$23.56	$23.81
Saturn 16th	$27.41	$27.50	$27.58
Saturn 18th	$24.91	$25.00	$25.08
Saturn 20th	$22.91	$23.00	$23.08
Neptune 16th	$28.00	$28.02	$28.03
Neptune 18th	$25.50	$25.52	$25.53
Neptune 20th	$23.50	$23.52	$23.53
Pluto 16th	$27.13	$27.14	$27.16
Pluto 18th	$24.63	$24.64	$24.66
Pluto 20th	$22.63	$22.64	$22.66

Currency Futures (Canadian Dollar, Australian Dollar) CS=1/1024, N=6,8,9

PLANET & HARMONIC	JAN 1	JULY 1	DEC 30
Jupiter 6th	$0.93	$0.95	$0.96
Jupiter 8th	$0.78	$0.80	$0.81
Jupiter 9th	$0.73	$0.75	$0.76
Saturn 6th	$0.91	$0.91	$0.92
Saturn 8th	$0.76	$0.76	$0.77
Saturn 9th	$0.71	$0.71	$0.72
Neptune 6th	$0.94	$0.94	$0.95
Neptune 8th	$0.79	$0.79	$0.80
Neptune 9th	$0.74	$0.74	$0.75
Pluto 6th	$0.89	$0.89	$0.89
Pluto 8th	$0.74	$0.74	$0.74
Pluto 9th	$0.69	$0.69	$0.69

Currency Futures (Euro and British Pound) CS=1/512, N=5,6,8

PLANET & HARMONIC	JAN 1	JULY 1	DEC 30
Jupiter 6th	$1.38	$1.42	$1.45
Jupiter 8th	$1.26	$1.30	$1.33
Jupiter 9th	$1.11	$1.15	$1.18
Saturn 6th	$1.33	$1.35	$1.35
Saturn 8th	$1.21	$1.23	$1.23
Saturn 9th	$1.06	$1.08	$1.08
Neptune 6th	$1.41	$1.41	$1.41
Neptune 8th	$1.29	$1.29	$1.29
Neptune 9th	$1.14	$1.14	$1.14
Pluto 6th	$1.30	$1.30	$1.30
Pluto 8th	$1.18	$1.18	$1.18
Pluto 9th	$1.03	$1.03	$1.03

Wheat and Corn Futures CS=1/64, N=2,4,8

PLANET & HARMONIC	JAN 1	JULY 1	DEC 30
Jupiter 2nd	$7.10	$7.36	$7.61
Jupiter 4th	$6.20	$6.46	$6.71
Jupiter 8th	$5.75	$6.01	$6.26
Saturn 2nd	$6.71	$6.80	$6.88
Saturn 4th	$5.81	$5.90	$5.98
Saturn 8th	$5.36	$5.45	$5.53
Neptune 2nd	$7.30	$7.32	$7.33
Neptune 4th	$6.40	$6.42	$6.43
Neptune 8th	$5.95	$5.97	$5.98
Pluto 2nd	$6.43	$6.44	$6.46
Pluto 4th	$5.53	$5.54	$5.56
Pluto 8th	$5.08	$5.09	$5.11

Soybean Futures CS=1/32, N=2,4,8

PLANET & HARMONIC	JAN 1	JULY 1	DEC 30
Jupiter 2nd	$12.39	$12.93	$13.43
Jupiter 4th	$11.49	$12.03	$12.53
Jupiter 8th	$11.04	$11.58	$12.08
Saturn 2nd	$11.61	$11.80	$11.96
Saturn 4th	$10.71	$10.90	$11.06
Saturn 8th	$10.26	$10.45	$10.61
Neptune 2nd	$12.80	$12.83	$12.86
Neptune 4th	$11.90	$11.93	$11.96
Neptune 8th	$11.45	$11.48	$11.51
Pluto 2nd	$11.05	$11.08	$11.11
Pluto 4th	$10.15	$10.18	$10.21
Pluto 8th	$9.70	$9.73	$9.76

Crude Oil Futures

PLANET & HARMONIC	JAN 1	JULY 1	DEC 30
Jupiter 1st	$78.38	$80.50	$82.50
Jupiter 2nd	$60.38	$62.50	$64.50
Jupiter 3rd	$54.38	$56.50	$58.50
Saturn 1st	$75.25	$76.00	$76.63
Saturn 2nd	$57.25	$58.00	$58.63
Saturn 3rd	$51.25	$52.00	$52.63
Neptune 1st	$80.00	$80.13	$80.25
Neptune 2nd	$62.00	$62.13	$62.25
Neptune 3rd	$56.00	$56.13	$56.25
Pluto 1st	$73.00	$73.13	$73.25
Pluto 2nd	$55.00	$55.13	$55.25
Pluto 3rd	$49.00	$49.13	$49.25

CHAPTER FIFTEEN

Conclusion

I have taken you on a wide-ranging journey in this Almanac to acquaint you with the mathematical and astrological links between planetary activity and market price behavior. I sincerely hope you will embrace financial astrology as a valuable tool to assist you in your trading and investing activity. I hope you will pause often to contemplate whether the correlations you have learned about in this Almanac are the actions of the cosmos on the emotions of traders and investors or the actions of a select few power players using astrology to manipulate the markets.

If you decide to embrace financial astrology as a tool to help you navigate the markets, I encourage you to stick with it. At first it might seem daunting, but fight the urge to give up. Soon enough, it will all make sense and your trading and investing activity will take on a new meaning.

To encourage you to stick with it and master the use of astrology, I will leave you with the words of Neil Turok from his 2012 book, *The Universe Within.* [1]

"Perseverance leads to enlightenment. And the truth is more beautiful than your wildest dreams."

NOTES & RECOMMENDED READING

Introduction

1) McWhirter, L. (1938) *McWhirter Theory of Stock Market Forecasting*. Astro Book Company, USA.
2) Bradley, D. (1948) Stock Market Prediction. Llewellyn Publishers, USA.

Chapter 1

Figure 1: supercoloring website (2015) [online] Available at: **http://www.supercoloring.com/coloring-pages/solar-system-model-worksheet**

Figure 2: bhavanajagat.com website (2017) [online] Available at: **https://bhavanajagat.com/2017/01/14/makar-sankranti-saturday-january-14-2017-perception-of-changing-seasons/**

Figure 3: elsaelsa.com (2011) The Glyphs for the Zodiac Signs.

[online]Available at: **https://www.elsaelsa.com/astrology/ zodiac-sign-glyphs.**

Figure 4: earther-rise.com website (2021) [online] Available at: **http://earther-rise.com/interpretation-guide/planet-glyphs**

Figure 5: physics.unlv.com website (2021) [online] available at: **http://www.physics.unlv.edu/~jeffery/astro/ial/ial_003. html**

Figure 8: ohio state astronomy 161 website (2021) [online] Available at: **http://www.astronomy.ohio-state.edu/~pogge/ Ast161/Unit2**

Chapter 3

1) McWhirter, L. (1938) *McWhirter Theory of Stock Market Forecasting.* Astro Book Company, USA.

Chapter 4

1) Braden,G. (2009) *Fractal Time.* Hay House, USA.
2) Haulman, C. (2010) The Panic of 1819. [online] Available at: **https://www.moaf.org/exhibits/checks_balances/ andrew-jackson/materials/Panic_of_1819.pdf. Accessed: November 2021.**
3) Richardson, G., Sablik, T. (2015) Banking Panics of the Guilded Age. [online] Available at: **https://www. federalreservehistory.org/essays/banking-panics-of-the- gilded-age**. Accessed: November 2021.
4) Reinhart, C., Rogoff, K. (2009). *This Time is Different: Eight Centuries of Financial Folly.* Princeton University Press.
5) Moen, J., Tallman, E. (2015) The Panic of 1907. [online] Available at: **https://www.federalreservehistory.org/ essays/panic-of-1907**. Accessed: November 2021.
6) NASA Eclipse website (2021) Eclipses and the Saros. [online] Available at: **https://eclipse.gsfc.nasa.gov/SEsaros/**

SEsaros.html. Accessed: November 2021.

7) NBER website (2021) US Business Cycle Expansions and Contraction. [online] Available at: **https://www.nber.org/ research/data/us-business-cycle-expansions-and-contractions**. Accessed: November 2021.

8) Vernon, J. R. (1991). The 1920-21 Deflation: The Role of Aggregate Supply. *Economic Inquiry*. 29 (3),pp: 572–580.

Chapter 6

1) Takahashi, F., Shimizu, H., Tsunakawa, H. (2019) Mercury's anomalous magnetic field caused by a symmetry-breaking self-regulating dynamo. *Nature Communications*. 10 (208).

Figure 16: ohio state astronomy 161 website (2021) [online] **http://www.astronomy.ohio-state.edu/~pogge/Ast161/ Unit2**

Chapter 8

1) Cahn, J. (2011) *The Harbinger*. Charisma Media, USA.
2) Cahn, J. (2016) *The Book of Mysteries*. Charisma Media, USA.
3) Cahn, J. (2017) *The Paradigm*. Charisma Media, USA.

Chapter 9

1) Gann, W.D. (1927) *Tunnel Through the Air*. Pantainos Classics, USA.
2) Lawlor, G. (1982) *Sacred Geometry*. Thames and Hudson, USA.

Chapter 10

1) Kramer, J. (1995) *Astrology Really Works*. Hay House, USA.
2) Gann, W.D. (1927) *Tunnel Through the Air*. Pantainos Classics, USA.

Chapter 11

1) McWhirter, L. (1938) McWhirter *Theory of Stock Market Forecasting.* Astro Book Company, USA.

Chapter 13

1) Gann, W.D. (1927) *Tunnel Through the Air.* Pantainos Classics, USA.
2) NASS website (2019). [online] Available at: **https://www. nass.usda.gov/Publications/Todays_Reports/reports/ cropan19.pdf**. Accessed: November 2021.
3) NASS website (2020) [online] Available at: **https://www. nass.usda.gov/Publications/Todays_Reports/reports/ cropan20.pdf**. Accessed: November 2021.

Chapter 14

1) Ferrera, D. (2015) *Gann for the Active Trader.* Cosmological Economics, USA.
2) Oreste, F. (2011) *Quantum Trading.* J. Wiley & Sons, USA.

Chapter 15

1) Turok, N. (2012) *The Universe Within.* House of Anansi Press, Canada.

Recommended Reading

The Bull, the Bear and the Planets, M.G. Bucholtz, (USA, 2013)

The Lost Science, M.G. Bucholtz, (USA, 2013)

Stock Market Forecasting – The McWhirter Method De-Mystified, M.G. Bucholtz, (Canada, 2014)

The Cosmic Clock, M.G. Bucholtz, (Canada, 2016)

The Universal Clock, J. Long, (USA, 1995)

A Theory of Continuous Planet Interaction, *NCGR Research Journal*, T. Waterfall, Volume 4, Spring 2014, pp 67-87.

Financial Astrology, Giacomo Albano, (U.K., 2011)

GLOSSARY

Ascendant: one of four cardinal points on a horoscope, the Ascendant is situated in the East.

Aspect: the angular relationship between two planets measured in degrees.

Autumnal Equinox: (see Equinox) – that time of year when Sun is at 0 degrees Libra.

Conjunct: an angular relationship of 0 degrees between two planets.

Cosmo-biology: changes in human emotion caused by changes in cosmic energy.

Descendant: one of four cardinal points on a horoscope, the Descendant is situated in the West.

Ephemeris: a daily tabular compilation of planetary and lunar positions.

Equinox: an event occurring twice annually, an equinox event marks the time when the tilt of the Earth's axis is neither toward or away from the Sun.

First Trade chart: a zodiac chart depicting the positions of the planets at the time a company's stock or a commodity future commenced trading on a recognized financial exchange.

First Trade date: the date a stock or commodity futures contract first began trading on a recognized exchange.

Full Moon: from a vantage point situated on Earth, when the Moon is seen to be 180 degrees to the Sun.

Geocentric Astrology: that version of Astrology in which the vantage point for determining planetary aspects is the Earth.

Heliocentric Astrology: that version of Astrology in which the vantage point for determining planetary aspects is the Sun.

House: a $1/12^{th}$ portion of the zodiac. Portions are not necessarily equal depending on the mathematical formula used to calculate the divisions.

Lunar Eclipse: a lunar eclipse occurs when the Sun, Earth, and Moon are aligned exactly, or very closely so, with the Earth in the middle. The Earth blocks the Sun's rays from striking the Moon.

Lunar Month: (see Synodic Month).

Lunation: (see New Moon).

Mid-Heaven: one of four cardinal points on a horoscope, the Mid-Heaven is situated in the South.

New Moon: from a vantage point situated on Earth, when the Moon is seen to be 0 degrees to the Sun.

North Node of Moon: the intersection points between the Moon's plane and Earth's ecliptic are termed the North and South nodes. Astrologers tend to focus on the North node and Ephemeris tables clearly list the zodiacal position of the North Node for each calendar day.

Orb: the amount of flexibility or tolerance given to an aspect.

Retrograde motion: the apparent backwards motion of a planet through the zodiac signs when viewed from a vantage point on Earth.

Sidereal Month: the Moon orbits Earth with a slightly elliptical pattern in approximately 27.3 days, relative to a fixed frame of reference.

Sidereal Orbital Period: the time required for a planet to make one full orbit of the Sun as viewed from a fixed vantage point on the Sun.

Siderograph: a mathematical equation developed by astrologer Donald Bradley in 1946 (By plotting the output of the equation against date, inflection points can be seen on the plotted curve. It is at these inflection points that human emotion is most apt to change resulting in a trend change on the Dow Jones or S&P 500 Index).

Solar Eclipse: a solar eclipse occurs when the Moon passes between the Sun and Earth and fully or partially blocks the Sun.

Solstice: occurring twice annually, a solstice event marks the time when the Sun reaches its highest or lowest altitude above the horizon at noon.

Synodic Month: during a sidereal month (see Sidereal Month), Earth will revolve part way around the Sun thus making the average apparent time between one New Moon and the next New Moon longer than the sidereal month at approximately 29.5 days. This 29.5 day time span is called a Synodic Month or sometimes a Lunar Month.

Synodic Orbital Period: the time required for a planet to make one full orbit of the Sun as viewed from a fixed vantage point on Earth.

Vernal Equinox: that time of the year when Sun is at 0 degrees Aries.

Zodiac: an imaginary band encircling the 360 degrees of the planetary system divided into twelve equal portions of 30 degrees each.

Zodiac Wheel: a circular image broken into 12 portions of 30 degrees each. Each portion represents a different astrological sign.

ABOUT THE AUTHOR

Malcolm Bucholtz, B.Sc, MBA is a graduate of Queen's University (Faculty of Engineering) in Canada and Heriot Watt University in Scotland (where he received an MBA degree and a M.Sc. degree). After working in Canadian industry for far too many years, Malcolm followed his passion for the financial markets by becoming an Investment Advisor/Commodity Trading Advisor with an independent brokerage firm in western Canada. Today, he resides in Saskatchewan, Canada where he trades the financial markets using technical chart analysis, esoteric mathematics and the astrological principles outlined in this book.

Malcolm is the author of several books. His first book, *The Bull, the Bear and the Planets*, offers the reader an introduction to financial astrology and makes the case that there are esoteric and astrological phenomena that influence the financial markets. His second book, *The Lost Science*, takes the reader on a deeper journey into planetary events and unique mathematical phenomena that influence financial markets. His third book, *De-Mystifying the McWhirter Theory of Stock Market Forecasting* seeks to simplify and illustrate the McWhirter methodology. Malcolm has been writing the *Financial Astrology Almanac* each year since 2014.

Malcolm maintains a website (www.investingsuccess.ca) where he provides traders and investors with astrological insights into the financial markets. He also offers the *Astrology Letter* service where subscribers receive twice-monthly previews of pending astrological events that stand to influence markets.

OTHER BOOKS
BY THE AUTHOR

The Bull, The Bear and The Planets

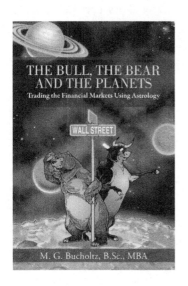

Once maligned by many, the subject of financial astrology is now experiencing a revival as traders and investors seek deeper insight into the forces that move the financial markets.

The markets are a dynamic entity fueled by many factors, some of which we can easily comprehend, some of which are esoteric. *The Bull, The Bear and the Planets* introduces the reader to the notion that astrological phenomena can influence price action on financial markets and create trend changes across both short and longer term time horizons. From an introduction to the historical basics behind astrology through to an examination of lunar astrology and planetary aspects, the numerous illustrated examples in this book will introduce the reader the power of astrology and its impact on both equity markets and commodity futures markets.

The Lost Science

The financial markets are a reflection of the psychological emotions of traders and investors. These emotions ebb and flow in harmony with the forces of nature.

Scientific techniques and phenomena such as square root mathematics, the Golden Mean, the Golden Sequence, lunar events, planetary transits and planetary aspects have been used by civilizations dating as far back as the ancient Egyptians in order to comprehend the forces of nature.

The emotions of traders and investors can be seen to fluctuate in accordance with these forces of nature. Lunar events can be seen to align with trend changes on financial markets. Significant market cycles can be seen to align with planetary transits and aspects. Price patterns on stocks, commodity futures and market indices can be seen to conform to square root and Golden Mean mathematics.

In the early years of the 20th century the most successful traders on Wall Street, including the venerable W.D. Gann, used these scientific techniques and phenomena to profit from the markets. However, over the ensuing decades as technology has advanced, the science has been lost.

The Lost Science acquaints the reader with an extensive range of astrological and mathematical phenomena. From the Golden Mean and Fibonacci Sequence, to planetary transit lines and square roots through to an examination of lunar Astrology and planetary aspects, the numerous illustrated examples in this book will show the reader how these unique scientific phenomena impact the financial markets.

Stock Market forecasting:
The McWhirter Method De-Mystified

Stock Market Forecasting -
The McWhirter Method
De-Mystified

M.G. Bucholtz, B.Sc., MBA

Very little is known about Louise McWhirter, except that in 1937 she wrote the book *McWhirter Theory of Stock Market Forecasting*.

In my travels to places as far away as the British Library in London, England to research financial Astrology, not once did I come across any other books by her. Not once did I find any other book from her era that even mentioned her name. I find all of this to be deeply mysterious. Whoever she was, she wrote only one book. It is a powerful one that is as accurate today as it was back in 1937. The purpose of writing this book is suggested by the title itself – to de-mystify McWhirter's methodology.

The Cosmic Clock

THE COSMIC CLOCK
TIMING THE FINANCIAL
MARKETS USING
THE PLANETS
M.G. Bucholtz, B.Sc, MBA

Can the movements of the Moon affect the stock market?

Are price swings on Crude Oil, Soybeans, the British pound and other financial instruments a reflection of planetary placements?

The answer to these questions is YES. Changes in price trends on the markets are in fact related to our changing emotions. Our emotions, in turn, are impacted by the changing events in our cosmos.

In the early part of the 20th century, many successful traders on Wall Street, including the venerable W.D. Gann and the mysterious Louise McWhirter, understood that emotion was linked to the forces of the cosmos. They used astrological events and esoteric mathematics to predict changes in price trend and to profit from the markets.

However, by the latter part of the 20th century, the investment community has become more comfortable in relying on academic financial theory and the opinions of colorful television media personalities, all wrapped up in a buy and hold mentality.

The Cosmic Clock has been written for traders and investors who are seeking to gain an understanding of the cosmic forces that influence emotion and the financial markets.

This book will acquaint you with an extensive range of astrological and mathematical phenomena. From the Golden Mean and Fibonacci Sequence through planetary transit lines, quantum lines, the McWhirter method, planetary conjunctions and market cycles. The numerous illustrated examples in this book will show you how these unique phenomena can deepen your understanding of the financial markets and make you a better trader and investor.

Made in United States
North Haven, CT
25 January 2022

15275433R00115